9/11

*The Lying Brain*

 *The Lying Brain*

LIE DETECTION
*in*
SCIENCE *and* SCIENCE FICTION

Melissa M. Littlefield

*The University of Michigan Press* / *Ann Arbor*

Copyright © by the University of Michigan 2011
All rights reserved
Published in the United States of America by
The University of Michigan Press
Manufactured in the United States of America
⊚ Printed on acid-free paper

2014   2013   2012   2011      4   3   2   1

*A CIP catalog record for this book is available from the British Library.*

Library of Congress Cataloging-in-Publication Data

Littlefield, Melissa M., 1979–
     The lying brain : lie detection in science and science fiction /
Melissa M. Littlefield.
        p.   cm.
     Includes bibliographical references and index.
     ISBN 978-0-472-07148-7 (cloth : alk. paper) — ISBN 978-0-472-
05148-9 (pbk. : alk. paper) — ISBN 978-0-472-02702-6 (e-book)
     1. Lie detectors and detection.   I. Title.
HV8078.L58      2011
177'.3—dc22                                          2010047503

*To* ZACH *and* SPENCER,
*for playing along*

The true story is multiple and vicious and untrue.
—MARGARET ATWOOD, "True Stories" (1981)

# Acknowledgments

I am indebted to many who have generously given of their time and attention to aspects of this project: to Susan Squier who encouraged this book from the first archival find to the final stages of writing; to my many writing group friends who provided constructive criticism over the years: Megan Brown, Jenell Johnson, John Marsh, Elizabeth Mazzolini, Justine Murison, Marika Seigel, and Shannon Walters; to Hugh Egan, Gus Perialas, and Joseph Tempesta for getting me into this business; and to those who read various drafts of the manuscript: Mark Morrisson, Janet Lyon, Philip Jenkins, Robert Mitchell, Bruno Clarke, Ron Schleifer, Bob Markley, and Debbie Hawhee. I would also like to thank Elizabeth Wilson, and members of the Society for Literature, Science, and the Arts, who listened to and commented on early versions of chapters 1, 3, and 5; and Ken Alder for sharing a copy of William Marston's dissertation as well as conversations about deception. My thanks to Thomas Dwyer for his enthusiasm about this project and to Alexa Ducsay, Christina Milton, and the production and editorial staff at the University of Michigan Press for their excellent work on the book. Jessica Mercado meticulously checked several chapters of the manuscript for errors; any remaining mistakes or omissions are my own. My gratitude to John Littlefield, Valerie Perialas, Kris Perialas, and Lori Eframson for encouraging my habits of writing, reading, and searching through dusty books; to Jonathan Littlefield and Jen Crabtree for seeing the humor in all things literary; to Doris Perialas and Alex Perialas for reminding me to love the work I do; and to my grandfather, John Perialas, who inspired me to ask the questions. Finally, I am forever grateful for Zach Sosnoff, who grew up with this manuscript; and I am indebted to Spencer Schaffner, for always knowing just what to say.

An earlier version of chapter 5 appeared in *Science, Technology & Human Values;* I am grateful to SAGE/SOCIETY for permission to reprint it here. The final, definitive version of "Constructing the Organ of Deceit: The Rhetoric of fMRI and Brain Fingerprinting in Post-9/11 America" has been published in *Science, Technology & Human Values* 34, May 2009, by SAGE Publications Ltd./SAGE Publications, Inc., All rights reserved.

# Contents

# Introduction

In 2008, the *Harvard Business Review* listed fMRI lie detection among its "Breakthrough Ideas." The inclusion of this brain-imaging technology is somewhat surprising given that neither the technology nor its application to deception is new: functional magnetic resonance imaging (fMRI) was developed in the early 1990s as a way to monitor brain functions; almost immediately, the technology was employed in laboratory experiments that attempted to detect deception in the brain. What have changed over the past decade are American investments in lie detection. From commercial applications, including No Lie MRI (located in Nevada), to Homeland Security, lie detection has once again[1] gained cultural cachet as a promising, progressive technology, this time via neuroscience and the panic of post–9/11 anxiety. After the terrorist attacks on the World Trade Center in 2001, U.S. government officials, scientists, and independent researchers began to subsidize and popularize technologies that they hoped would more accurately and quickly detect threats to national security.

What the *Harvard Business Review*—and a host of recent academic and popular articles on fMRI lie detection—takes for granted is that fMRI represents real progress in the battle against deception. In the recent past, Paul Root Wolpe and Daniel Langleben, coauthors of the review essay and psychiatric colleagues at the University of Pennsylvania, have disagreed about the promise of fMRI lie detection. Wolpe, a bioethicist, has penned many worried pages about the potential dangers of the latest applications of brain-imaging technologies; Langleben, one of the first to spearhead an experimental initiative involving fMRI and lie detection, has been a savvy but wary advocate of the technology's application. What the two researchers agree on is the progress being made through fMRI lie detection.

> Until recently, we had not improved very much on the methods of the ancient Greeks, who took the pulse of a suspect under questioning— a rudimentary polygraph in concept. But recent research using functional magnetic resonance imaging, or fMRI, has begun to identify the areas of the brain involved in deception. These laboratory experiments (many done by coauthor Daniel D. Langleben) suggest that accurate, reliable lie detection is finally within reach. (2008, 25)

A quick glance at such a statement might bring a sense of relief: at long last a solution to what Wolpe and Langleben describe as the "ubiquitous, yet difficult to detect" problem of deceit. However, Wolpe and Langleben's narrative of technological and scientific progress covers over a much more dynamic genealogy of approaches to lie detection and begs several questions: Why should they compare the brain-imaging technology of fMRI to the polygraph, to the taking of a pulse, to ancient methods of interrogation? Indeed, why leap from the Greeks to the twenty-first century in the space of one clause? Why assume that laboratory experiments are a recent (if not recently successful) phenomenon in the science of lie detection? Why insinuate that the brain's blood oxygenation levels, which is what fMRI lie detection relies on, are a better means for the detection of deception than the taking of the body's pulse?

This book begins by questioning what bioethicists, scientists, pundits, and the press often take for granted: that lie detection research has remained relatively unsophisticated until its most recent leap into brain-imaging technologies and that forms of mechanical lie detection preceding brain-based technologies can be subsumed under the rubric of "polygraphy," the process of graphically recording several of the body's autonomic responses to various stimuli. Through impoverished and de-historicized representations, an artifact (the polygraph) and an imagined machine (the lie detector) have come to stand in for a century of technological explorations and expectations about deception and its mechanical detection.

*The Lying Brain* argues that instead of providing a break from or a novel approach to lie detection, neuroscience is implicitly recycling scientific and cultural assumptions about deception and its mechanical detection: that lies are measurable phenomena that manifest themselves in the body's physiology, particularly in the autonomic and central nervous systems; that the body produces objective data that are easily and unam-biguously interpretable; and that deception demands the knowledge and suppression of truth. These assumptions took shape during the first half of the twentieth century and were the result of a dialogue between scientists from psychology, physiology, and forensics; law enforcement agents; lawyers; and authors of science fiction and scientific detective fiction. Literature produced from these dialogues (scientific publica-tions, government documents, journalism, and fiction) provides insights into the formation of assumptions about deception and its mechanical detection.

By highlighting the mutual imbrication of literature and science in the genealogy of lie detection technologies, I position deception and its mechanical detection within the American cultural imagination—a space that includes not only what Hugo Münsterberg, the founder of American applied psychology, termed the "wider tribunal of the general reader" (1908, 10–11) but also the scientific laboratory. Just as the basic theoretical assumptions of lie detection have remained relatively stable over the past century, so too has the patterned repetition of appeals to progress and benevolence. Lie detection has maintained its grip on the American cultural imagination because its proponents always promise something better: better interrogation methods, a more civilized police force, better crime control, a better society, better knowledge about other minds, and better justice. The astute marketing of these promises in science and science fiction has kept a potentially marginal technology at the forefront of American culture. In recent decades, the ahistoricist impulse of the neurosciences has continued to maintain the promises and accept the underlying assumptions of lie detection technologies.[2]

## Lie Detection's Foundational Assumptions

Before going any further, let me say a few words about the assumptions I have identified here. First, mechanical lie detection assumes that decep-tion can be defined by and is synonymous with its effects on the body, so much so that scientists interpret specific sets of physiological reactions as indicative of deception. The polygraph, for example, measures auto-nomic nervous functions of the body (heart rate, blood pressure, and respiration), in an attempt to trace the emotional charge—and there-fore the potential deception—behind any statement. Functional mag-netic resonance imaging (fMRI) assumes that increased blood oxygenation levels in the brain are indicative of brain activity in that region;[3] researchers claim to have identified several areas that appear to be more active whenever we attempt deception. In each case, deception is defined within the parameters of a laboratory experiment, which means that deception is a controlled variable; it can be contrasted with another known variable, usually described as the truth; and, within these parameters, deception requires additional work on the part of the subject, which can be read as and from changes in physiology.

In *The Lying Brain* I focus on this laboratory-based definition of deception and return to it specifically in chapters 2 and 5; however, it would be a mistake to imagine that deception is merely a laboratory construct, or that the laboratory is isolated from either sociohistorical conceptions of deception or implications for lie detection deployment by police, lawyers, employers, and the American government. Nevertheless, in their experimental design(s), production of knowledge(s), and representation(s) of those knowledges to entities outside of the laboratory, scientists play a crucial role in simplifying and consolidating answers to a set of complex questions: Does deception require intentionality? Can the term encompass activities from concealment to falsification to fabrication? Is deception the opposite of truth? Can deception be nonverbal? Is it emotional? Must it be successful? Is deception about escaping consequences or seeking rewards? Is deception an evolutionary adaptation? Is it something only humans can do?[4]

Instead of further complicating these questions, definitions of deception illustrate a certain (and sometimes unspoken) consensus about deception's core components. In one of the most recent and summative studies of deception's definition in neuroscientific literature, Masip, Garrido, and Herrero (2004) point out that many research teams rely on at least one of the three assumptions identified by Coleman and Kay's study of the prototypical lie (1981): a lie "is characterized by (a) falsehood, which is (b) deliberate and (c) intended to deceive" (28).[5] Masip, Garrido, and Herrero go on to offer the following definition of deception: "Deception can be understood as the deliberate attempt, whether successful or not, to conceal, fabricate, and/or manipulate in any other way factual and/or emotional information, by verbal and/or nonverbal means, in order to create or maintain in another or in others a belief that the communicator himself or herself considers false" (2004, 148). The problem with this definition is not its scope, which is impressive, but its application. While experiments can create environments in which subjects are deliberate and manipulative, the key question is how to measure intentionality: whether we have identified, isolated, and measured "a belief that the communicator himself or herself considers false." William Marston is one of the first to address this problem through his conception of the deceptive consciousness, which I discuss in chapter 2.

For the remainder of *The Lying Brain,* I define deception not as an act but as an object—what Loraine Daston and Peter Galison characterize as a "working object" and define as "any manageable communal representatives of the sector of nature under investigation" (1992, 85). Working objects are active and dynamic; they can change over time and between locations. They represent a way to talk about scientific objects as both material and immaterial, as real and constructed. Daston and Galison's definition accounts for the real/material act of deception and the elaborate definitional construction of that act within particular spaces, specifically the laboratory and its representative publications. Within this theoretical framework, the first assumption on which lie detection is based—that a set of physiological changes accompany a deceptive act—becomes an attempt to capture a working object that exceeds both graphic records and semiotics, even as it is partially captured by each.

When scientists measure deception by and through changes in physiology, they depend upon the second (and related) assumption: that the body provides us with objective data that do not require interpretation; or put another way, the body appears to be self-reporting. The body speaks for itself in large part because the aspects of physiology being measured are not under direct, conscious control. So, whether scientists measure blood pressure, respiration, electrical skin conductance, heart rate, blood oxygenation levels, or electrical activity, the body seems to provide information regardless of the subject's conscious thought or desire. The paradox is that even as the body speaks, what it is saying requires interpretation, often by scientific experts. As Bruno Latour, Steve Woolgar, and Donna Haraway (among others) have argued, the reading of bodies and other objects is not innocent or transparent; at best, bodily reading is translational. The political, economic, and ideological agendas foundational to science affect which bodies or objects will be "read" and how they will be best interpreted. Bruno Latour explains this paradox and the power of scientists to speak *for* and *through* their objects in *We Have Never Been Modern* (1993), noting with plenty of irony that "scientists declare that they themselves are not speaking; rather, facts speak for themselves. . . . Little groups of gentlemen take testimony from natural forces, and they testify to each other that they are not betraying but translating the silent behavior of objects" (29). As I illustrate in each chapter, the translation from physiological change to diagnosis of deception requires a large interpretive leap.

The idealization of the self-reporting body is practically and figuratively foundational not only to lie detection but also to the sibling forensic technology of fingerprinting that I discuss in chapter 4. Practically, both lie detection and fingerprinting construct the body as the litmus test for truth about a person. Each technology seeks what Francis Galton termed the self-signatures of the body—those invariable markers of individuality (fingerprinting) or deception. Metaphorically, the individuality of the fingerprint is marshaled by another graphic, brain-based lie detection technology, Brain Fingerprinting (a registered trademark of Brain Fingerprinting Laboratories), which measures the electrical activity of the brain via EEG. Brain Fingerprinting is said to produce a catalog of the information and experiences stored in one's memory, a catalog that is as unique as one's fingerprint.

Finally, lie detection depends on the often unspoken assumption that deception is the known suppression of truth. Beginning in the early twentieth century and extending to contemporary brain-based lie detection, experimental protocols have taken for granted that when we lie we must know and suppress the truth. In contemporary parlance, the truth has been characterized as more biologically efficient than deception, because deception is said to expend more energy. So, tracing a genealogy of lie detection necessitates tracing a genealogy of truth's representation as natural and foundational. This is not to reiterate an old philosophy-of-science argument that laboratory science accesses Truth through its experiments; rather, it is a claim about the ways that laboratory experiments are both products and producers of cultural conceptions of truth and its relation to deception. I take up this assumption more specifically in chapters 2, 3, and 5 as I examine William Marston's laboratory experiments concerning the deceptive consciousness, the literacy of mind reading, and discourses surrounding contemporary laboratory lie detection using fMRI and EEG.

Each of this book's five chapters contextualizes the three foundational assumptions of lie detection via the sociohistorical development and deployment of scientific theories, literatures, and debates that continue to bear on technologies and practices of lie detection. The first two chapters address theories coined before either the polygraph or the lie detector was conceived; they include Hugo Münsterberg's psychological principle that the hidden feeling will betray itself (circa 1907) and William Marston's theory of the deceptive consciousness (1913–22). In chapter 3, I explore the development of mind reading between 1930 and 1950 as a kind of preliteracy for brain imaging. And, in the final chapters of the book, I turn to the recent uptake of older concepts, including the trope of "fingerprinting" as a marker for individuation in biometrics, DNA, and Brain Fingerprinting; and the paired concepts of the biological mind and biological truth. Each concept was and remains foundational to the functioning of contemporary neuroscientific assumptions about deception, its mechanical detection, and the bodies on and through which these technologies work.

Importantly, the assumptions behind and the concepts deployed in service of lie detection were not developed or popularized by science alone. In each case, various forms of popular fiction and journalism played a significant role in at least publicizing, if not predicting or cocrafting, lie detection. Muckraking scientific detective fiction such as Edwin Balmer and William MacHarg's *The Achievements of Luther Trant* (serialized in 1909/published as a book in 1910); popular bestsellers such as Alfred Bester's *The Demolished Man* (1952/1953), Jack Finney's *Invasion of the Body Snatchers* (1954/1955), and James Halperin's *The Truth Machine* (1996); even fictional alibis created in William Marston's Harvard laboratory experiments, all capture key debates of their eras in a timely dialogue with science. Although canonical literature is important to the discussion of lie detection and informs studies such as Ronald Thomas's *Detective Fiction and the Rise of Forensic Science* (1999), popular and noncanonical literatures have earned academic cachet over the past decades. When scholars look to little magazines (Morrison 2001), manifestoes (Lyon 1999), pulp detective fiction (Smith 2000), and pulp science fiction stories (Squier 2004) to explain modernist culture and biocultures, they provide us with a more complete and complex portrait of the social milieu at any given moment. In *The Lying Brain,* I value these alternate, popular literatures, because in and through these stories, we find marketing strategies for deception-detection technologies; lessons in the literacy of mechanized mind reading; and once-imaginary, now-deployed connections between the brain, the mind, and deception.[6]

In a similar vein, each chapter addresses larger discussions concerning reactions to lie detection among several key constituencies: journalists, lawyers, police, and government officials. From the *Frye v. U.S.* case that decided the fate of lie detection in American courts for over seventy years, to J. Edgar Hoover's fetishization of the forensic sciences in the 1950s, to post–9/11 funding for fMRI and EEG lie detection, *The Lying Brain* addresses the problematic and often schismatic character of lie detection at each stage of development, deployment, and popularization. By working within the matrix of scientific theories, literatures, and cultural debates, *The Lying Brain* not only situates neuroscientific lie detection in a century-long history, but also illustrates the importance of various literatures to the survival of lie detection as a technology of the American cultural imagination.

## Key Terms

Throughout the book, I use three terms, which are related but not interchangeable: lie detection, *the* lie detector, and the polygraph. I define and use the term *lie detection* to describe mechanical methods for measuring and recording acts of deception. Whereas interrogators once used a mouthful of rice, boiling water, or hot coals to determine deception (and/or guilt), lie detection's postindustrial form involves tracking and graphically recording changes in the body's autonomic and central nervous systems so as to correlate these changes with various emotions and events, such as deception.

Lie detection has become a largely American phenomenon (Alder 2007), but the emotional inscription technologies on which it relies have a long international history.[7] Some of the first men to experiment with lie detection did so in Italy and Germany. Cesare Lombroso, a criminologist, employed a plethysmograph to record the relationship between the volume of fluid in a subject's limb and a subject's deceptive responses in 1895. Lombroso speculated that a drop in volume indicated deceit. In 1914, Vittorio Benussi used a pneumograph to test subjects' respiration for similar correlations between deception and changes in the autonomic nervous system. Hugo Münsterberg, who immigrated to the United States in 1892, began a laboratory research program to study deception at Harvard. Between 1913 and 1925, Münsterberg's student, William Marston, worked with a sphygmomanometer to measure correlations between what he termed the "deceptive consciousness" (Marston 1917, 153) and blood pressure.

These instruments—the plethysmograph, the pneumograph, and the sphygmomanometer—were eventually combined to create the first poly-graph, literally a "multiple" "graph." Thus, *the polygraph* is not just any machine, it is a specific artifact: a technology developed first by John Larson in 1921 and later modified by Leonarde Keeler in 1926.[8] As I explain in chapter 1, the polygraph was and continues to be a controversial technology that, nonetheless, is still widely applied. From the Lindberg baby kidnapping in 1932, to the screening of German POWs in the 1940s, to McCarthy's anticommunist campaign of the 1950s, to the high-profile (criminal) cases of the 1990s—including JonBenet Ramsey's murder and the Aldrich Ames spy case—the polygraph has become a staple of American criminal investigation (Alder 2002, 2007).

The polygraph continues to be a feature of the public's imagination, particularly when it is conflated with *the* lie detector, a term coined by journalists between 1917 and 1918, and based on William Marston's publicized experiments with a single instrument, the sphygmomanometer. One of the earliest references to the lie detector can be found in the February 17, 1918, edition of the *Chicago Daily Tribune*, which was published shortly after William Marston's first paper on the systolic blood pressure test for deception (1917). The technology featured in the newspaper's "Camp Stories Contest" bears little resemblance to emotional inscription technologies; however, the power of lie detection lies not in its authorized technologies but in its mythical ability to secure a confession. In this particular story, rife with racialized images that will reappear in chapters 1, 4, and 5, a "Negro" man is exposed as a thief: "this is a lie detector," explains a lieutenant, "if you are telling a lie this hand will point at you when I loosen this wheel" (B6). As newspapers began reporting the results of his research in actual and embellished prose, William Marston was careful to distance himself from *the* lie detector, noting on several occasions that he was not the inventor of the lie detector but the creator of "the lie detector test" (Marston 1938), a protocol for interrogation using a sphygmomanometer. I therefore define *the lie detector* not as an artifact but as an imagined instrument, an accumulation of the lore, desires, hopes, and dreams of the scientific, journalistic, and lay communities. This so-called lie detector has been developed in and through various hypotheses, publications, and demonstrations. Much of the controversy surrounding the practices of lie detection can be linked back to both the applications of the polygraph and the imagined potential of the lie detector.[9]

## Beyond Good or Bad Science

Whether we speak of the lie detector, the polygraph, or fMRI lie detection, one question generally overrides all others: Does lie detection work? Polygraphers and analysts, ethicists and lawyers often assume that the answer to this question lies in a refutation or confirmation of the instruments' and techniques' accuracy; they have yet to reach a consensus. This question is often better translated as, Is lie detection good science or is it just bad/junk science? One of the problems with this question, with a distinction between "bad" and "good" science, is that it retains "good" science as not only possible but also linked to ideal conceptualizations of objectivity and value neutrality. Positing practices as "bad" or "good" science presumes that better protocols, an adherence to (what Sandra Harding termed "weak") objectivity, and attention to the scientific method could correct "bad" science. Instead, as many feminist science studies scholars have argued, we could stop talking about "bad" science as subjective, invested, situated, and imbricated with/in culture, and start recognizing that all science—even, and especially, science that appears to be most objective—is always a social, cultural, contextual product; it is hardly value neutral (Keller and Longino 1986; Harding 1991; Haraway 1991).

Because I support this deconstruction of "good" and "bad" science, I am less interested in whether or not lie detection technologies "work" (whether they are good or bad science) and more interested in examining the ways that lie detection has maintained its hold on the American cultural imagination *despite* its track record of failures and scandals. As I have already indicated, one piece of this puzzle is the progressive narrative that adheres to lie detection's early instruments *and* contemporary brain-based detectors: it does not matter whether lie detection works, so long as it provides the promise of changing society for the better. Likewise, the genealogy I trace in *The Lying Brain* illustrates that technologies do not have to work to successfully accomplish their stated task. One could argue that the ultimate purpose of lie detection is not identifying but eliminating deception through the threat of detection. This theme emerges in William Marston's scientific and popular writing, in science fiction from *The Demolished Man* to *The Truth Machine,* and in marketing materials for brain-based lie detection. In *The Truth Machine,* for example, as lie detection becomes a socially ubiquitous technology, people simply stop lying.

By looking beyond the question of efficacy, and at a combination of literatures, *The Lying Brain*'s approach, methodology, and cultural analyses are unique among scholarship on the polygraph and the lie detector, which has, thus far, been distributed among lawyers, police officers, psychologists, historians, engineers, and a handful of cultural theorists.[10] Historian Ken Alder has reconstructed the early history of the great men of polygraphy, including William Marston, John Larson, and Leonarde Keeler, while also exploring the larger contexts of our American obsession with the lie detector. Psychologist Geoffrey Bunn argues that the media has rendered the lie detector—its tests, techniques, results, and uses—meaningful when and where the science of the era could not. Closest to my concerns, Ronald Thomas has included detective fiction in the cultural equation to argue that the forensic sciences (including lie detection), the figure of the detective, and the genre of detective fiction emerged at a cultural moment rife with questions and assumptions about "devices of truth" that purport to allow the body to speak for itself (1999, 10).

Methodologically, cultural theorists have adopted a Foucauldian approach that characterizes the deceptive body as interpretable and integrated within specific discourses. Thomas, for example, has argued that detective fiction aided in the nineteenth-century shift from early conceptions of character to modern notions of identity as detectives sought to create narratives about the criminal based on information gleaned from his and his victims' bodies: "The systematic medicalization of crime in criminological discourse during this period corresponded to the literary detective's development into a kind of master diagnostician, an expert capable of reading the symptoms of criminal pathology in the individual body and the social body as well" (1999, 3). The new literacy Thomas tracks is a useful way to understand the purpose of each graphic, bodily trace. Indeed, *The Lying Brain* takes up the literacy and disciplinary literatures that seek to render meaningful the measurements of the body necessitated by lie detection of all kinds (chapters 3 and 4).

My approach emphasizes the centrality of literature in any history of lie detection, while also intervening in literature and science studies to incorporate the significance of a third term: technology.[11] Technology is conspicuously absent from paradigms such as "literature and science" and typically reserved for the history of science and technology or Science, Technology, and Society (STS) scholarship. The introduction of this third term helps to constitute a shift away from a dichotomous "two cultures" model and allows me to triangulate the relationship between literature and science with a genealogy of mechanical lie detection.

## Introducing Technology to Literature and Science

Whereas scholars once spoke in terms of the "two cultures" to describe the problematic relationship between the humanities and the sciences, the field has now moved to a more productive and less antagonistic model that can be described as the multiple cultures (Shusterman 1998) or "cultural matrix" (Thiher 2005) model. This approach does not pit literature and science against each other, nor does it attempt a simple reversal that would replace the primacy of science with the primacy of literature. Instead, a multiple cultures or cultural matrix model accounts for the differences of time, space, disciplinary development, and boundaries that inform the rise of and ostensible divisions—and relations—among science, technology, fiction, and cultural media. *The Lying Brain* takes as one of its first propositions that literature and science are neither commensurable nor incommensurable; they cannot simply be "read" through the language of the other, nor are they as utterly incompatible as the C. P. Snow and F. R. Leavis debate insinuated. As recent scholars have noted, literature and science scholarship could argue that, at the very least, literature refracts science (Thiher 2005) or that literature is a "remodeling or remediation of scientific representation" (Clarke and Henderson 2002, 7) and that literature can help science estrange itself from its conventions, practices, and language (Dimock and Wald 2007).

I would go one step further to urge us to think beyond refraction, beyond remediation, beyond even estrangement to consider the mutuality of science and literature in the shaping of technology. First, the addition of technology as a third term is a reminder that science and literature are cultural products that are often engaged in similar pursuits, working in concert with each other, for example, to further the development and/or marketing of a particular technology such as the polygraph or even *the* lie detector. In the case of lie detection, popular fiction was and remains a key component in the marketing of and the creation of an imagined lay vision for the devices emerging from physiological and psychological laboratories. From the first decade of the twentieth century to the last, American literature explored the implications of lie detection technologies, both actual and imagined.

However, this particular triad of literature, science, and technology is not simply about literature commenting on science or marketing already finished technologies; indeed, fiction has also made calls for cross-disciplinary applications of lie detection while also serving as an early predictor of the turn to the brain as the telltale organ that can be probed and measured to uncover deception. In *The Achievements of Luther Trant* (1909/1910), for example, instruments such as the plethysmograph, the pneumograph, and the sphygmomanometer were featured as the latest law enforcement tools long before their acceptance by police or in courts of law. *The Truth Machine* (1996) predicts that a technology akin to fMRI lie detection will become a ubiquitous, indispensable, and commercial component of modern life—this, long before companies such as No Lie MRI emerged.

The second reason to add technology to the literature and science dyad is that technologies demonstrate interdisciplinarity whenever they combine existing materials in new ways. As Otniel Dror has cogently argued, lie detection was a "new application and innovative interpretation" of established emotional-inscription instruments (1999b, 376); likewise, the polygraph was constructed of several physiological instruments that had been adopted in psychological laboratories before being deployed to effect change in law enforcement, criminal interrogation, and judicial practice. Examining this genesis and diffusion provides a fresh perspective on the interdisciplinary work performed by literature, science, and technology including the mixing of genres, techniques, methods, and their distribution among various audiences. One of the earliest genres of fiction to take up lie detection was scientific detective fiction, itself a conglomeration of science fiction, detective fiction, and muckraking journalism. As I demonstrate in chapter 1, the success of these stories is intimately tied to another piece of literature, Hugo Münsterberg's *On the Witness Stand* (1908), a series of essays on the utility of lie detection for courts of law. By including technology as a third term, interrelations between and within technological, generic, and scientific histories become apparent.

Third, technologies and theorizations of technologies in the context of literature and science remind us to think beyond the bane-or-boon model. Technologies are neither good nor evil; rather they are employed, deployed, dynamic, and often patterned. Instead of asking questions about the inherent value of technology, *The Lying Brain* addresses how lie detection has been put to use as a means to enact larger sociocultural changes. In chapter 1, for example, I argue that the publication of literature about early lie detection instruments was a strategy to both inform and endear these technologies to the public. Through this campaign, which stretched over twenty years, instruments of lie detection were marshaled as the solution to problematically violent police interrogations and unsolvable crimes.

Finally, expanding a literature and science methodology to include technology also allows for fruitful discussions between literary theorists and technology scholars (whether in STS or history of science and technology) who are already analyzing many of the genealogical threads relevant to a cultural history of lie detection. Robert Brain, Joseph Dumit, Anne Beaulieu, and Otniel Dror have theorized the history of the graphic trace, the meanings and impacts of imaging technologies such as positron-emission tomography (PET scans) for definitions of personhood, the history and implications of brain mapping, and the history of emotional inscription technologies respectively. None of these scholars, however, speaks to or about the fictional narratives that emerged concurrently with the technologies they study. Such narratives not only help us to understand the larger cultural milieu from and into which each of these technologies emerged, but they also provide a sense of how various disciplines, media, and publics were being shaped by, prepared for, and receptive to the ideologies that undergird emergent technologies. In short, scholars in STS and literature and science are doing their respective jobs, but in so doing, they are only telling part of the story. In each of its five chapters, *The Lying Brain* knits together the fields of literature and science and STS through the shared term of technology.

In chapter 1, "Selling the Psychological Detective: Hugo Münsterberg's Applied Psychology and *The Achievements of Luther Trant, 1907–30*," I argue that early failures for lie detection in American courts should be framed by a much larger public championing of lie detection in science, popular media, and fiction. To make my case, I trace the genealogy and uptake of mechanical lie detection via the development, dissemination, and reception of Hugo Münsterberg's principle: that the hidden feeling will betray itself. Münsterberg, William Marston's mentor, had a profound effect on the history of lie detection in part because he was willing to seek and receive aid from the popular press. My analysis is framed by the paired circulation of Münsterberg's essay collection, *On the Witness Stand* (1908), and a set of fictional stories about a psychological detective, *The Achievements of Luther Trant* (1910a). Both texts were initially published— and eventually republished—within two years of each other in muckraking publications. Throughout the formative and transitional years of the early twentieth century, the "achievements" of Luther Trant perform several key functions as they voluntarily respond to Hugo Münsterberg's call for "a wider tribunal," including the validation of lie detection.

In chapter 2, "The Science of Lying in a Laboratory: William Marston's Deceptive Consciousness, 1913–22," I analyze Marston's concept of the deceptive consciousness with particular attention to the foundational assumptions informing its definition and use in laboratory experiments. Marston's measurement techniques, developed between 1913 and 1922, emerge from and help to redefine a long history of emotional inscription technologies. Tracing the generation and history of the deceptive consciousness also provides a bridge between the early literary and legal marketing of lie detection from *The Achievements of Luther Trant* and Hugo Münsterberg's applied experimental psychology work to current brain-based experiments on deception using fMRI and EEG. In all cases, scientists are looking to find "the marks of crime on men." As a theory, Marston's concept of the deceptive consciousness provides a concrete means through which to examine the experimental assumptions that undergird this relatively modern pursuit of fragmenting the body into discrete units for analysis.

In chapter 3, "Thought in Translation: Reading the Mind in Science and Science Fiction, 1930–50," I argue that contemporary representations of brain imaging in the media rely on the trope of mechanical mind reading not as a hyperbole but as a literacy. One aspect of this literacy can be traced back to science and science fiction between the 1930s and the 1950s, an era in which several scientific theories conspired to make thought-energy matter. From EEG to ESP, to the thought translators of science fiction, thought was understood as an energy form that could become visible through various scientific and technological interventions. I trace four narratives that emerged during this era and that continue to inform the brain imaging technologies I discuss in chapter 5: thought is energy; thought energy can be transparently translated (via machine) into sounds and images; primitive and dangerous thoughts reside in the hidden self and lower brain; and finally, thought can be used as evidence of intention in criminal justice cases. To make my argument, I examine scientific texts on telepathy and ESP by René Warcollier, Upton Sinclair, and J. B. Rhine alongside Hans Berger's early experiments on human electroencephalography (EEG) and science fiction in which thought is made visible as picture, sound, and/or text. The most prominent of these narratives is Alfred Bester's *The Demolished*

*Man* from 1953, which, as we will see in the coda, is later copied and modified by James Halperin in his 1996 novel about brain-based detection, *The Truth Machine.*

In my fourth chapter, "Without a Trace: Brain Fingerprinting, Biometrics, and Body Snatching," I turn away from lie detection instruments per se and toward the sibling forensic technology of fingerprinting. Like the lie detector, fingerprinting depends on and reinforces ideologies of the self-reporting body—that body that ostensibly divulges objective information about the subject. Under this rubric, deceptive bodies can be ferreted out through the use of scientific techniques. By examining scientific, political, and fictional representations of the fingerprint between 1892 and 1954, I argue that the fingerprint has become a code for individuation, one that has been applied to any number of technologies from DNA fingerprinting to Brain Fingerprinting. Cultural representations of fingerprinting, including Jack Finney's *Invasion of the Body Snatchers* (1955), examine how the internalization of fingerprint as code reveals the coconstitution of classificatory system and object.

Chapter 5, "A Tremor in the Brain: fMRI Lie Detection, Brain Fingerprinting, and the Organ of Deceit in Post–9/11 America," engages contemporary systemic representations of lie detection and identification. The EEG-based technology of Brain Fingerprinting and fMRI have been hailed as the next, best technologies for lie detection in America, particularly in a post–9/11 context. In scientific journals and the popular press each has been juxtaposed with, and deemed superior to, traditional polygraphy, which measures changes in the autonomic nervous system and correlates these fluctuations with emotions such as anxiety, fear, and guilt. Brain Fingerprinting and fMRI detection ostensibly go beyond the autonomic body to focus on the central nervous system, require less subjective interpretation, and enable the visualization of deception's biological basis. I contend that the juxtaposition of polygraphy and brain-based detection is an argumentative strategy that foregrounds the corrective advantage of brain-based techniques, even as it creates an artificial rupture between contiguous technologies and ignores the shared assumptions foundational to fMRI, EEG, and the "truth telling" technologies of polygraphy and fingerprinting. Far from describing the brain and its functions, fMRI and Brain Fingerprinting produce models of the brain that reinforce social notions of deception, truth, and deviance.

Finally, the "Coda: Lie Detection as Patterned Repetition—From *The*

*Demolished Man* to *The Truth Machine*" focuses on the recognition of patterns that emerge throughout the history of lie detection and in each of this book's chapters. Here, I challenge the progressive narrative of scientific advancement by returning to a reiterative model that seeks commonalities and repetitions. To make my argument, I compare the treatment of lie detection technologies in two novels I have discussed elsewhere in the book: Alfred Bester's *The Demolished Man* and James Halperin's *The Truth Machine*. Both books imagine a future world in which a new lie detection technology has been systemically implemented as a means to improve human relationships and reduce crime. Both books also imagine a moment when the system is undermined by a protagonist who commits murder and escapes arrest by repeating a poem whenever he is interrogated. While each system is eventually reinstated, the reauthorization of lie detection necessitates that the protagonist undergo a radical brain reshaping procedure that neutralizes his threat as murderer and hacker. Commenting on these common themes, problems, and solutions can illustrate the genealogical origins and patterned repetitions of lie detection that continue to accrue in contemporary neuroscientific practice.

Taken together these chapters function as case histories in a genealogy of lie detection that refuses both an easy opposition between literature and science and a progressive narrative of technological advancement. As the chapters illustrate, literature and science have often worked in tandem to develop literacies, disseminate information, market technologies, and even predict technological successes and/or failures. Each of these efforts, whether cooperative or discrete, depends on the triangulation of science, literature, and technology. Here, the technologies of lie detection serve to complicate the relationship between science and literature.

In their plurality and ideological assumptions, these same technologies of lie detection also problematize a progressive linear narrative that begins with the polygraph. While the *Harvard Business Review* counts contemporary brain-based detection among the "Breakthrough Ideas for 2008," the arguments made throughout this book call that status into question by offering a more complex genealogy that accounts for century-long similarities—rather than ruptures—between historical and contemporary versions of deception detectors. To begin this discussion, I turn to one of the foundational assumptions of lie detection: that hidden emotions can be made visible via measurements of the body.

# 1

## Selling the Psychological Detective

Hugo Münsterberg's Applied Psychology and
*The Achievements of Luther Trant, 1907–30*

In his 1908 collection of essays, *On the Witness Stand: Essays on Psychology and Crime,* Hugo Münsterberg expounds upon one of the principles that would inform lie detection for the next century: "the hidden feeling betrays itself" (113).[1]

> It may be easy to suppress intentionally the conspicuous movements by which we usually accentuate the emotions. It is not necessary to become wild with anger and to collapse in sorrow, we may even inhibit laughter and tears. . . . But the lips and hands and arms and legs, which are under our control, are never the only witnesses to the drama which goes on inside—if they keep silent, others will speak. The poets know it well. (114)

Münsterberg, a German émigré, student of Wilhelm Wundt, and founder of American applied psychology, believed that emotions exceed our conscious control; that they affect not only our psychology, but also our physiology; and that they could, therefore, be measured and made legibly useful to other fields, including law enforcement and courts of law.

In the same revelatory breath that explains his primary principle, Münsterberg commends not the scientists that came before him, but the poets, those authors who represent subtle signs of emotion in the bodies of their characters: "There is hardly a tragedy of Shakespeare in which the involuntary signs of secret excitement do not play their role. . . . The helpless stammering of the excited lover betrays everything which his deliberate words are to deny" (114–15). Münsterberg's recognition of the poets is neither accidental nor naive; as Otniel Dror notes, "the participants in and developers of the new science [of emotional inscription] were not oblivious to the competing technologies of poets, writers, painters, and actors who shared in the quest for representation. They did not reject these alternative knowledge makers off-hand, but attempted to enlist their representations for scientists' own ends" (1999b, 368). Indeed, Münsterberg's record of scholarship and popular writing allies his own work to that of the poets. However, Münsterberg may not have realized that his descriptive statement was also prophetic: over the course of the century, "the poets"—those writers who developed, discussed, and/or disseminated applied psychology's notions about measurable emotion through narrative—would play an increasingly important role in the deployment and marketing of one particular kind of applied psychology: lie detection technologies, techniques, and principles.

At about the same time that Münsterberg was writing and collecting his essays on crime, law, and psychology, two Chicago newspapermen, Edwin Balmer and William MacHarg, began to compose the adventures of a new kind of hero: a psychologist-detective named Luther Trant. First published in *Hampton's Magazine* in 1909 and 1910, and later collected as *The Achievements of Luther Trant* (1910), the Luther Trant stories embodied Münsterberg's principle and its application to crime through the use of various instruments for the detection of deception. Remarkably, the dialogue between Münsterberg's principle and Balmer and MacHarg's fiction continued to inform the marketability of lie detection well into the 1920s and 1930s, even—and especially—after Münsterberg's protégé, William Marston, failed to find legal acceptance for his lie detection test (which used a sphygmomanometer)[2] in the landmark *Frye v. U.S.* case of 1923. After this failure to find juridical authorization, which I will discuss later in this chapter, Münsterberg's collected essays and a series of Luther Trant stories were republished in an attempt to secure public acceptance for mechanical lie detection. The latter were republished by Hugo Gernsback in *Amazing Stories* and *Scientific Detective Monthly* (1926–30).

Despite their popularity—and *strategic republication*—the Luther Trant stories have been overlooked[3] by scholars working on the cultural history of lie detection and the history of the polygraph.[4] This is, perhaps, because very often the technologies used in these stories have been seen as proto/pseudoscience or characterized as emotion inscription technologies (Dror 1999b) differentiated from the polygraph or even *the* lie detector.[5] In what follows, I construct an alternate genealogy of lie detection that is informed by Münsterberg's principle and framed by the paired circulation of Münsterberg's essays and Luther Trant's adventures. Throughout the early decades of the twentieth century, Münsterberg, Balmer, MacHarg, and later William Marston and Hugo Gernsback were invested in marketing applied psychology as a progressive correction to corruption in law and police work. Indeed, the work of each figure was the product of an era marked by several key debates: the development of psychology as a discipline distinct from philosophy and physiology, the relevance and relationship of psychology to police work and the law, the change in status of witness testimony, and the continuing professionalization of the police force in America.

I first situate both *On the Witness Stand* and *The Achievements of Luther Trant* in the initial struggle to apply psychological techniques and principles to matters of law and police work between 1907 and 1910. Next, I discuss the legal invalidation of Marton's systolic blood pressure test for deception—and by extension mechanized lie detection—through *Frye v. United States* (1923). Finally, I suggest that the popular authorization of lie detection technologies post-*Frye* was further aided by the paired circulation of Münsterberg's republished edition of *On the Witness Stand* (1925) and Hugo Gernsback's fiction magazines *Amazing Stories* and *Scientific Detective Monthly* (1926–30), which recirculated several Luther Trant stories. My goal is to explain, via the span of a thirty-year period, the mutual imbrications of literature and science in the conception and distributed of lie detection as an applied psychological technology.

## Hugo Münsterberg's "Wider Tribunal," 1907–10

The longer a discipline can develop itself under the single influence, the search for pure truth, the more solid will be its foundations. But now experimental psychology has reached a stage at which it seems natural and sound to give attention also to its possible service for the practical needs of life.

—HUGO MÜNSTERBERG (1908, 8)

When Hugo Münsterberg accepted William James's invitation to direct Harvard's psychology laboratory in 1892, he arrived at a site of disciplinary (re)formation as the burgeoning discipline of American psychology was working to differentiate itself from both philosophy and physiology. While Münsterberg embraced the new direction of his primary discipline (he was a founding member of the American Psychological Association, and his ideas about the mind and its relation to the body were aligned with the American psychologists from the functionalist and behaviorist schools), he also served as president of the American Philosophical Association in 1908, and his psychological laboratory at Harvard had all the trappings of a physiology laboratory.

In his laboratory practices and theoretical outlook, Münsterberg attempted to authorize psychology as a natural science through instrumentation. For him, as for many psychologists of the day, "the success of psychology as both an experimental science and form of applied knowledge was predicated upon its ability to replicate the mathematical precision and predictive validity found in the natural sciences" (Ward 2002, 111). One of Münsterberg's hypotheses, which dovetailed with the work of several of his contemporaries, was that physiological changes can be correlated to mental and emotional states such as guilt, fear, joy, excitement, anger, and relief.[6] Because Münsterberg's primary goal was the application of psychology to other fields, he was particularly interested in inspecting the minds of criminal suspects and courtroom witnesses by measuring changes in their body's physiology. He tested his hypothesis using instruments that originated in various physiological laboratories, which could measure changes in the autonomic nervous system: the sphygmomanometer to monitor changes in blood pressure, the pneumograph to track changes in the frequency and depth of respiration, the automatograph to measure muscle contraction, and the plethysmograph to track changes in the volume of blood in a limb.

Among Münsterberg's practical and metaphorical favorites was the chronoscope,[7] used to measure a subject's reaction time to various stimuli, particularly during word-association tests.[8] He used several types of chronoscopes in his laboratory,[9] including the Hipp chronoscope, which required the subject to hold a lip key in their mouth. A small screen was then dropped in front of the subject to start the timed test; this screen typically had a single word written on it. The test ended when the subject either read the word provided aloud or provided a word that s/he associate with the word in front of them. As s/he spoke, the lip key dropped from his/her mouth, stopping the chronoscope from recording time.

Rhetorically, the chronoscope came to embody Münsterberg's ideas about the measurable relationship between body and mind and the instrumental connections between applied psychological and the natural sciences, as described in *On the Witness Stand.*

> The chronoscope of the modern psychologist has become, and will become more and more, for the student of crime what the microscope is for the student of disease. It makes visible that which remains otherwise invisible, and shows minute facts which allow a clear diagnosis. The physician needs his magnifier to find out whether there are tubercles in the sputum: the legal psychologist may in the future use his *mental microscope* to make sure whether there are lies in the mind of the suspect. (1908, 77; my emphasis)

Although the chronoscope simply records the time between events, Münsterberg's comparison construes the instrument as both optical and invasive: a "mental microscope." Such terminology begins to establish the literacy of mind reading I discuss in chapter 3, by suggesting the transparency of mind via instrumentation. Such insinuations jibe with other depictions of emotional inscription technologies in the late nineteenth and early twentieth centuries. Indeed, Münsterberg's characterization is both a product and producer of a shift in ways of seeing,[10] as sensorial experience was being supplanted by mechanistic observation. The X-ray, the microscope, the chronoscope, along with a host of other technologies, produced representations of bodily processes that "differed significantly from previous artistic depiction in their mode of production, form and style of representation, method of interpretation, and use" (Dror 1999b, 360). This instrumental sight also challenged divisions between public and private, internal and external, physiology and psychology, the body and its emotions (Ward 2002; Thomas 1999; Dror 1999b).

And yet, as Otniel Dror notes, and Münsterberg's description of the chronoscope illustrates, there is an intriguing contrast "between the simplicity of many of these instruments and their mythical power to 'dive into . . . minds'" (Dror 1990b, 364). In the passage from *On the Witness Stand,* for example, as the chronoscope is metaphorically transformed into a "mental microscope," it becomes "a magnifying-glass for the most subtle mental mechanism, and by it the secrets of the criminal mind may be unveiled" (Münsterberg 1908, 108). Distinctions between the autonomic nervous system, the brain, the mind, the emotions—and lies—are collapsed in a Münsterbergian mythology of visualization and access that we will see again in chapter 3. Münsterberg's specific reference to finding "lies in the mind of the suspect" begins to construct another myth, that of *the* lie detector, a machine that can distinguish between deception and truth by measuring the body. In the rest of this chapter, we will see how the fictional psychologist-detective Luther Trant translates the gist of Münsterberg's mythos into an applied task for the expert: finding "the marks of crime on men's minds."

Despite his somewhat zealous rhetoric in *On the Witness Stand,* Münsterberg does make an important distinction that challenges several older myths about criminal types: he chooses lies, and not liars, as his proper object of study. In contrast to scholarship on lie detection thus far, it should be noted that Münsterberg is interested in lies as discrete phenomena that are not necessarily associated with any one type of person;[11] put another way, he does not discuss "the liar" as a "human kind" (Bunn 1997, 101). Because Münsterberg believes that lies are distinguishable objects within any mind, he points to the suspect not as type but as a storehouse for what he actually seeks: lies. These objects, he argues, can be best located through instrumentation that renders thoughts visible.[12]

His focus on lies brings us to the second half of Münsterberg's objective: to bring European notions about applied psychology to the American academy and lay population, including criminal investigations and the courtroom.[13] He believed "education, medicine, art, economics, and law" (Münsterberg 1908, 9) could equally benefit from "the new psychology" (20):[14] this blend of qualitative techniques and quantitative instruments said to reveal and record the inner workings of the mind, or as Münsterberg terms it, "the drama which goes on inside" (114).[15] Münsterberg argued that psychological experiments could translate to improved job performance and enhanced familial and civic relationships. Leading the way would be the student of psychology, whose

> experiments can indicate best how the energies of mill-hands can reach the best results, and how advertisements ought to be shaped, and what belongs to ideal salesmanship. And experience shows that the politician who wants to know and to master minds, the naturalist who needs to use his mind in the service of discovery, the officer who wants to keep up discipline, and the minister who wants to open minds to inspiration—all are ready to see that certain chapters of Applied Psychology are sources of help and strength for them. (10)

Absent from this laundry list of converts to applied psychology, which Münsterberg included in his 1908 collection, *On the Witness Stand,* are the lawyer and the judge.

Although most fields were open to psychology's influence, lawyers, judges, and even police officers were the most vocal opponents of applied psychology and its constituent instruments. Lawyers and legal scholars were particularly resistant to the psychologist as expert witness and arbiter of witness testimony (Blumenthal 2002), due, in part, to the fact that by the late nineteenth century, lawyers had finally taken precedence in the courtroom over witnesses whose testimony had been declared subjective and therefore problematic (Thomas 1999, 34). Police were wary of the implementation of psychological instruments during interrogations for fear that their authority would be undermined by the psychological expert. In Münsterberg's opinion, both lawyers and police acted like Luddites in their resistance to technological change: the police continued to use what was colloquially known as the third degree[16] as a means to extract confessions; judges relied on their own common sense and observations.

To this resistance, Münsterberg posited science as the ultimate arbiter of the mind, memory, and even the difference between truth and lies: "Cannot science help us out? Cannot science determine with exactitude and safety that which is vague in the mere chance judgment of police officers?" (1908, 117). Münsterberg proposed that the psychologist could better regulate the "the treachery of human memory" (44) by bringing specific instruments and techniques to bear on issues of witness testimony, criminal interrogation, and confession.

> There is thus really no doubt that experimental psychology can furnish amply everything which the court demands: it can register objectively the symptoms of the emotions and make the observation thus independent of chance judgment, and, moreover, it can trace emotions through involuntary movements, breathing, pulse, and so on, where ordinary observation fails entirely. (131)

As evidenced here, one of Münsterberg's primary goals was to make psychology useful to other fields. In so doing, Münsterberg hoped both to raise the profile of psychology, cementing its boundaries and authority, and also to remedy a variety of social ills, including, for example, the barbarism of the police's third-degree interrogations.

When his ideas were met with reproach by police and legal scholars (Moore 1907; Wigmore 1909),[17] Münsterberg called upon "the wider tribunal of the general reader" (1908, 11) to validate his ideas. *On the Witness Stand: Essays on Psychology and Crime* represents Münsterberg's collected efforts to persuade the public of psychology's import for the legal and criminal justice systems. Indeed, the texts for the collection were drawn not from legal journals but from his many popular articles on police and legal reform that had been published in *McClure's, Reader's, Times,* and *Cosmopolitan* magazines between January 1907 and March 1908. In the introduction to the collection, Münsterberg notes,

> The lawyer alone is obdurate. . . . The lawyer and the judge and the juryman are sure that they do not need the experimental psychologist. If the time is ever to come when even the jurist is to show some concession to the spirit of modern psychology, public opinion will have to exert some pressure. Just in the line of the law it therefore seems necessary not to rely simply on the technical statements of scholarly treatises, but to carry the discussion in the most popular form possible before the wider tribunal of the general reader. (1908, 10–11)

Here, Münsterberg singles out the lawyer as the lone resister to the "spirit of modern psychology" and triangulates the debate by introducing the general public as, if not an arbiter, at least an empowered participant in the discussion.[18]

However, by calling upon the public, Münsterberg drew the ire of psychological colleagues—including William James—who disagreed about the merits of applying psychology to other fields. Münsterberg, on the other hand, was not a purist; he and his student, William Marston, represented a "particular type of public psychologist—a group that would continue to shape the reputation of psychological knowledge throughout the century" (Ward 2002, 147), who were, in fact, infamous for popularizing psychology in public settings. Münsterberg not only wrote articles for the popular press but also consulted for the film industry and set up mental testing booths at the Chicago World's Fair in 1893. William James derogatorily termed the latter a "Münsterbergian Circus" because it had the flavor of a side-show demonstration (Hale 1980, 97; Ward 2002, 142). But, as I will illustrate in the next section, Münsterberg's detractors could not stop the dissemination of his ideas.

### Selling Münsterberg's Principle: *The Achievements of Luther Trant,* 1909–10

Within one year of *On the Witness Stand*'s publication, two *Chicago Tribune* newsmen, Edwin Balmer and William MacHarg, answered Hugo Münsterberg's call for a wider tribunal by creating a fictional psychological detective, Luther Trant. Balmer and MacHarg, who served as reporters for the *Tribune* beginning in 1903 and 1898, respectively, were strangers to neither journalism nor literature: Balmer had recently published *Waylaid by Wireless* (1909); MacHarg authored several short pieces of fiction for the *Tribune,* including "A Christmas Fantasy" (Dec. 18, 1898) and "Mr. Dudd of Chicago" (June 25, 1899). Separately, together, and with other coauthors, the men published over twenty books in genres from war stories, to scientific detective fiction, to romances and science fiction.[19] Their real fame—particularly in the Midwest—came from the publication of three books set in Illinois and parts of Michigan: *The Achievements of Luther Trant* (1910a), *The Blind Man's Eyes* (1916), and *The Indian Drum* (1917) (Obuchowski 1995). The Luther Trant stories were set in Chicago, the city that would become the epicenter for lie detection research between 1920 and 1940.

Much like Sherlock Holmes, Trant is called upon to solve crimes, including embezzlement, murder, and espionage, without resorting to violence or using traditional weapons of any kind. What distinguishes the Luther Trant stories, and the subgenre of scientific detective fiction[20] to which they belong, from detective fiction is the application of instruments and principles from experimental psychology to gather information and interrogate suspects. Whereas Sherlock Holmes uses deduction and analyzes trace evidence to solve crimes, Trant relies on the application of the chronoscope, galvanometer, plethysmograph, sphygmograph, and pneumograph, along with principles akin to Münsterberg's "mental microscope" to find not the marks of crime on the environment but the marks of crime on men's minds. Indeed, the Luther Trant stories are the earliest American fiction to imagine the application and acceptance of experimental psychological instruments to forensic detective work. In so doing, they break new ground for detective fiction while remaining responsive to and reflective of the historical and then-contemporary debates about the relationship between psychology and the law.

The eleven[21] Luther Trant stories were first serially published in 1909 and 1910 in *Hampton's Magazine,* before being collected as *The Achievements of Luther Trant* in 1910.[22] Trant's first venue, *Hampton's,* was the intellectual and financial brainchild of Benjamin Hampton, who took over the magazine[23] in 1905 just after Theodore Dreiser became an editor for the publication. From its modest beginnings as a rejuvenated but failing publication, Dreiser and Hampton raised circulation to over 100,000 by 1907. After Dreiser left in 1907, *Hampton's* circulation continued to rise;

by the time of its financial crisis in 1911, circulation was up to 400,000. In this venue, the Luther Trant stories were enormously successful, so much so that they were often featured in the advertisements for the magazine.

Trant's stories fit well with Hampton and Dreiser's vision for *Hampton's* not only because of their exciting plots but also because of their muckraking attempts to expose the ineptitudes of police officers and a legal system that denied the importance of applied psychology. Indeed, Münsterberg's essays and Balmer and MacHarg's scientific detective fiction share an important affiliation with early twentieth-century muckraking: both published their work in magazines renowned for their journalistic efforts to expose corruption. "As early as 1893, [Münsterberg] had published with *McClure's,* but with the advent of muckraking he was called upon for more regular contributions" through which he "voiced a plea for penal reform" (Wilson 1970, 157). *Hampton's Magazine,* where Luther Trant first came to fame, has been characterized by Debi Unger and Irwin Unger as "an important muckraking journal of the day" (2005, 108). Both Münsterberg and the Chicago newsmen are invested in discourses of improvement and progress aimed at correcting the corruption of poor police work and legal blunders.

Luther Trant's narratives are insistent about the potential improvements that could be wrought by the broader application of new psychological technologies to police work and the legal system.[24] In particular, Trant is concerned with the "haphazard methods of the courts" (Balmer and MacHarg 1910a, 95), the torturous third-degree interrogations performed by police, and the basic inefficiency of criminal processing. According to Trant, the obstinate judicial system and police force need applied psychology in order to be more objective, more humane, and more effective. All of these agendas are made clear by the opening pages of "The Man in the Room," the first of the Luther Trant stories to be published. Throughout the lengthy preamble to the actual story (which concerns the death of a scientist in his laboratory) readers are privy to an animated discussion between Trant, the "brilliant, but hotheaded young aid" (1910a, 2), and his aging professor, Dr. Reiland. While the aging doctor is skeptical about seeing his techniques applied outside of psychology and even notes, "I, myself, am too old a man to try such new things" (5), Trant comes to represent the imminent sweeping changes brought by the "new psychology" (1910a, 325).[25]

Trant's first point is one of progress and humanity: "'It is astounding, incredible, disgraceful, after five thousand years of civilization our police and court procedures recognize no higher knowledge of men than the first Pharaoh put into practice in Egypt before the pyramids! . . . Five thousand years of being civilized,' Trant burst on, 'and we still have the third degree!" (Balmer and MacHarg 1910a, 1–2). For Trant, "civilization" necessarily implies progress, and in the case of psychology and interrogation, American courts and police have advanced very little. The third degree, a term used to characterize police interrogations that were physically and psychologically abusive, was prevalent and would remain so until the Wickersham Commission Reports of 1931.[26] Beyond the third degree, Trant faults the legal system for being unsystematic, unobjective, and inefficient. Citing several cases ripped from the headlines, Trant argues that through psychology "I shall not take eighteen months to solve [them]. I will not take a week" (1910a, 5). And not only will psychology be more efficient, it will also produce more reliable results: "there is no room for mistakes . . . in scientific psychology," Trant insists. "Instead of analyzing evidence by the haphazard methods of the courts, we can analyze it scientifically, exactly, incontrovertibly—we can select infallibly the true from the false" (95). Trant's new breed of detective is capable of modernizing criminal and legal institutions via civilized, scientized approaches to evidence.

In an effort to portray the realism of their psychological detective and his instruments to their readers, Balmer and MacHarg add an editorial foreword to their 1910 collection of Luther Trant stories. Sounding much like Hugo Münsterberg, and stressing the "factual" nature of their stories, the authors argue that

> if these facts are not used as yet except in the academic experiments of the psychological laboratories . . . it is not because they are incapable of wider use. . . .The hour is close at hand when they will be used not merely in the determination of guilt and innocence, but to establish in the courts the credibility of witnesses and the impartiality of jurors, and by employers to ascertain the fitness and particular abilities of their employees. (1910a, foreword)

Within the text, several lie detection techniques are even explored by Trant long before they are ever applied by polygraphers in actual criminal cases.[27] It is little wonder, then, that from the very first line of the collection, the authors are reticent to classify their collection as a work of the imagination. "Except for its characters and plot," Balmer and MacHarg explain in their foreword, "this book is not a work of the imagination." Implicit in this statement is the desire to disavow the imagination as dangerous to or orthogonal to scientific practice and validation—a phenomenon still visible in contemporary forensic textbooks.[28] Underlying this statement are several assumptions: that the imagination is unimportant, and potentially damaging, to the development and authorization of scientific technologies;[29] that literature needs to validate its own techniques by disavowing fictional foundations; and that science cannot successfully imagine its potential achievements through extrapolation. Ironically, what Balmer and MacHarg predict through their collection are the achievements of psychology for the forensic sciences, an exercise that demands imaginative thinking but not necessarily fictionalization.

Indeed, Trant's narratives imagine the as yet unaccomplished application of experimental psychology to criminal investigation and the courts, and thereby forge connections between psychology, the law, and the police that have, until this point, only been fantasies of applied psychology. By imagining the "achievements" of applied psychology, *The Achievements of Luther Trant* not only reflects Münsterberg's hopes but also illustrates the potential influence of fictional accounts on the development of scientific thought, experiment, and authorization. Or, as Leonard Krasner explains concerning the historical imbrication of fiction and psychology, "the fictional use of psychology not only illustrates an important application of psychology to the solution of crimes but also offers a portrait of the activity of psychologists to the very wide segment of the population that reads such books" (1983, 578). It makes perfect sense, then, that Trant's stories were, as one early reviewer noted, "absorbingly interesting to the student of psychology as well as to the general reader" (Display ad 9, no title, 1910, 12).

Luther Trant's liminal place (somewhere between fiction and nonfiction) both benefits, and benefits from, explicit and implicit references to Münsterberg. First, Luther Trant showcases the type of psychologist-detective that Münsterberg conceived but could not produce. He is a "one time assistant in a psychological laboratory, now turned detective" (1910a, Foreword) who uses psychological instruments to solve crimes. As a reviewer from the *New York Herald* notes more explicitly in an advertisement for the detective, Luther Trant is "a new style of detective. The basis of the new detective art and science is the use of measuring and recording instruments chiefly exploited heretofore by Professor Münsterberg" (Display ad 9, no title, 1910, 12).

Likewise, Münsterberg's basic assumptions about measurable connections between body and mind are regularly featured in the text. We are told in "The Hammering Man," for example, that "every emotion reacts upon the pulse, which strengthens in joy and weakens in sorrow, grows slower with anger, faster with despair; and as every slightest variation is detected and registered by the Sphygmograph" (Balmer and MacHarg 1910d, 715). Luther Trant also explains—in nearly the same language later used by Münsterberg's protégé William Marston—that the hidden feeling will betray itself: "No matter how hardened a man may be, no matter how impossible it may have become to detect his feelings in his face or bearing," argues Trant, "he cannot prevent the volume of blood in his hand from decreasing, and his breath from becoming different under the emotions of fear or guilt" (Balmer and MacHarg 1910a, 164);[30] or, as Marston argues in 1913, "no normal person can lie without effort. It is impossible to increase one's effort—mental, nervous, or otherwise—without increasing the strength of the heartbeat" (1938b, 29).[31] The presence of such surety maintains the believability of the fictional text while also reinforcing the authority of an as-yet-unauthorized science.

The instruments featured in Trant's stories all share basic foundational principles that originally linked them to Münsterberg and, by the 1920s and 1930s, connect them to the developing science of polygraphy. In "The Eleventh Hour," for example, as Trant revels in several of his past successes as a psychological detective, he acknowledges the instruments that Münsterberg himself championed.

> The delicate instruments of the laboratory—the chronoscopes, kymographs, plethysmographs, which made visible and recorded unerringly, unfalteringly, the most secret emotions of the heart and the hidden workings of the brain; the experimental investigations of Freud and Jung, of the German and French scientists, of Munsterberg and others in America—*had fired him with the belief in them and in himself.* (Balmer and MacHarg 1910a, 325; emphasis added)

The connection between Trant, Münsterberg, and their shared instruments, like the connection between literature, science, and technology, is multidirectional: the instruments create a synergy of belief that feeds both the scientific and popular imagination, which, in turn, creates space and authority for the machines. In the following sections we see that nearly all of the instruments referenced in the Trant collection were eventually taken up and modified by lawyers and police officers for the purpose of lie detection.

As we have already seen, the instruments, graphic records, and analyses cataloged in Trant's fictional tales participate in the shift in ways of seeing; they also participate in the mechanistic mythos that led Münsterberg to characterize the chronoscope as a "mental microscope." Thanks to Luther Trant, Münsterberg's "mental microscope" is transformed into a kind of literacy—a new language and strategy for reading—that we will reencounter in chapters 2 and 3. His instruments allow him to read the marks of crime on men's minds. As Trant explains it, the shift from Holmesian trace evidence to psychological measurement becomes obvious and elementary: "I read from the marks made upon minds by a crime," Trant explains, "not from scrawls and thumbprints upon paper" (Balmer and MacHarg 1910a, 88). In his opening debate with Professor Reiland, Luther Trant admonished his mentor to "teach any detective what you have taught to me, and if he has half the persistence in looking for the marks of crime on *men* that he had in tracing its marks on *things,* he can clear up half the cases that fill the jail in three days" (3). At least two things should be noted here: first, and most obviously, is Trant's assumption that psychological techniques could be applied to other fields through some basic education. Indeed, if police would only take up these new instruments, they could eliminate the third degree altogether in favor of a more "civilized" approach (3).[32] Second, and more important, is the insinuation that psychology can more accurately uncover criminality by examining the minds of men than by the mere examination of a crime scene. One such mark is guilt, detected in each narrative by the same instruments that Münsterberg championed: the galvanometer, sphygmograph, pneumograph, and plethysmograph.[33]

This new literacy is made more powerful and accessible because of the didacticism of Luther Trant's narratives. His adventures serve as a space to catalog, explain, and illustrate the interpretation of the graphic trace produced by psychologists and their instruments. On many occasions, these graphs are frequently heuristically simplified for the readership of *Hampton's Magazine.* "The Hammering Man," for example, includes the graphic output of the sphygmomanometer as an illustrative figure. The graphs represent the physiological reactions (in this case changes in the blood pressure) of three Russian revolutionaries as they listen to a young woman recount her father's brutal betrayal. The complex tracings of three different individuals over the course of a prolonged interview that should require an expert's interpretation are compressed into five lines of text. These are easily read by the expert psychological detective and even by the lay audience of the magazine. Although the graphs appear fairly similar, and all have been excerpted from any comparative coordinate system, Trant declares that "the test . . . has shown as conclusively and irrefutably as I could hope that that this man [Meyan] is not the revolutionist he claims to be, but is, as we suspected might be the case, an agent of the Russian secret police" (Balmer and MacHarg 1910d, 714). His conclusion is supported by a caption that aids a general reader's interpretation of the various spikes in the inscribed record.

> 1. Sphygmograph record of healthy pulse under normal conditions. 2 and 3 Sphygmograph records of Dmitri Vasili and Ivan Munikov when Eva Silber told of her father's betrayal; the lower and rapid pulsation thus recorded indicate grief and horror. 4. Record of Meyan on this occasion; the strong and bounding pulse indicates joy. 5. Meyan's sphygmograph record when Trant shows the yellow note that betrayed Herman Silber; the feeble, jerky pulse indicates sudden and overwhelming fear. (714)

From these graphs and their explanation, the reader learns several important lessons about mechanical lie detection: that there is such a thing as a "healthy" pulse and "normal" conditions under which physiology can be measured; that physiology can be equated with particular emotional states, as variable and specific as grief, horror, joy, and fear; that bodies react uncontrollably, but imperceptibly—at least to the naked human eye—when confronted with evidence implicating guilt or even, at the very least, recognition; and that, ultimately, all of these "facts" are perceptible by a machine and readable from a simple graph. An earlier story, "The Man Higher Up" (Oct. 1909), which concerns a corrupt shipping company president and the murder of a checker, also includes the graphic results of a test involving the sphygmomanometer and the plethysmograph. In addition to the pictorial representations, readers are provided with the following interpretation of their curves, an explanation that illustrates how a simple machine can produce a powerful mythos.

> If I had it here I would show you how complete, how merciless, is the evidence that you knew what was being done. I would show you how at the point marked 1 on the record your pulse and breathing quick-

ened with alarm under my suggestion; how at the point marked 2 your anxiety and fear increased; and how at 3, when the spring by which this cheating had been carried out was before your eyes, you betrayed yourself uncontrollably, unmistakably . . . how your pulse throbbed with terror; how, though unmoved to outward appearance, you caught your breath, and your laboring lungs struggled under the dread that your wrong doing was discovered and you would be branded—as I trust you will now be branded, Mr. Welter, when the evidence in this case and the testimony of those who witnessed my test are produced before a jury—a deliberate and scheming thief. (1910a, 183)

Implied here are assumptions not only about the instruments, their expert interpretation, and their ability to record emotion but also about their real, practical admissibility in courts of law, as we shall see in the next section concerning William Marston and the *Frye v. U.S.* case.

Ultimately, Luther Trant's proposed reform via instrumental ways of seeing and understanding the criminal, like Münsterberg's hypotheses, demands that psychology fashion that public following that can place pressure on police, lawyers, and judges. Thus, the final and equally crucial cumulative function of Trant's collected "achievements" is the production of a public following. By working with the instruments of lie detection, Trant's collected adventures directly address the issues of popular and legal acceptance of the lie detector decades before its admissibility was reviewed by American courts. Within the stories themselves, several references imagine and predict testing techniques and even crucial cultural centers of lie detection research. Before administering the tests in "The Man Higher Up," Trant explains the new psychological methods to Mr. Rentland—the U.S. Treasury spy who hired him—by referencing his own past cases: "I am a stranger to you, but if you have followed some of the latest criminal cases in Illinois perhaps you know that, using the methods of modern practical psychology, I have been able to get results where old ways have failed" (1910a, 162). Referring explicitly to his own achievements in Illinois, Trant connects himself to Chicago, the eventual center of lie detector investigation, which produced the Northwestern Scientific Crime Laboratory (1930), the publisher of the *American Journal of Police Science* (1930–32), and the workplace of Leonarde Keeler in 1929.

In "The Man Higher Up" in particular, Trant also confronts the contested space of the courtroom directly, long before the definitive verdict of *Frye* in 1923: "You are thinking now, I suppose, Mr. Welter . . . that such evidence as that directed against you cannot be got before a court. I am not so sure of that. But at least it can go before the public tomorrow morning in the papers, attested by the signatures of the scientific men who witnessed the test" (Balmer and MacHarg 1910a, 182). Crucial to this last statement is the dialogue between the spheres of the courtroom, the laboratory, and the public. Even if it is not accepted by the courts, the lie detector's testimony, in conjunction with the "signatures of the scientific men who witnessed the test," will be heard by the public.

As Trant successfully solves cases through the use of various instruments, he gains allies (including American and international corporations, private families, and the police) that reappear throughout later stories eager to laud the merits of Trant's new psychology. In fact, Trant's ability to convince even the greatest disbelievers—the police in particular—substantially precedes the actual acceptance of lie detection and other technologies by officers.[34] When he first meets Captain Crowley in "The Fast Watch," for example, the policeman is not only incredulous about Trant's "psycho" (1910a, 41) techniques but mocks him, calling him a "four-flushing patent palmist" (57). Yet, once Trant illustrates the physical manifestations of criminal guilt through the use of the galvanometer, Crowley and his lead investigator Walker have little choice but to accept his techniques. As in Münsterberg's own description of the Harvard laboratory, the inclusion of detailed instrumentation descriptions serves to reinforce the scientization of psychology, while also debunking the conflation of experimental psychology and parapsychology. Moreover, these same officers return to champion Trant in a later story, "The Empty Cartridges." When questioned about the validity of Trant's methods, Crowley himself replies, "Mr. Sheppard, it's myself has told you about Mr. Trant before; and I'll back anything he does to the limit, since I see him catch the Bronson murderer, as I just told you, by a one-cell battery that would not ring a door bell" (256). By the final story, "The Eleventh Hour," Trant has succeeded so well in applying his new psychology that he and his techniques are constantly in demand.

> As he hurried down Michigan Avenue now, he was considering how affairs had changed with him in the last six months. Then he had been a callow assistant in a psychological laboratory. The very professor whom he had served had smiled amusedly, almost derisively, when he had declared his belief in his own powers to apply the necromancy

of the new psychology to the detection of crime. . . . So well had he succeeded that now he could not leave his club even on a Sunday, without disappointing somewhere, in the great-pulsating city, an appeal to him for help in trouble. (1910a, 325)

Although this trajectory is also present in the individual stories as they were first published in *Hampton's Magazine,* their cumulative effect is even greater in the collected achievements of the psychological detective. What Münsterberg attempted, Trant completed: he challenged the doubters—the lawyers and police who denied the power of applied psychology—to prove the worth of the psychological detective and concretize the mythos of the machine. Even "in the face of misunderstanding and derision, [Trant] had tried to trace the criminal, not by the world-old method of the marks he had left on things, but by the evidence which the crime had left on the mind of the criminal himself" (Balmer and MacHarg 1910a, 325).

As Münsterberg urged and Balmer and MacHarg illustrated, lie detection achieved a position of relative prominence among police officers and the public—if not the courts—in the decades following the publication of *On the Witness Stand* and *The Achievements of Luther Trant*.[35] By predating and predicting the "achievements" of forensic science, *The Achievements of Luther Trant* illustrates the utility of fictional accounts: it predicts scientific advancement, aids in popularization, and influences the ways science can signify in culture. An analysis of Balmer's, MacHarg's, and Münsterberg's early collected work also reveals an alternative history for the development and dissemination of lie detection. We will see this even more clearly in the final section in which I specifically address several of Luther Trant's adventures that were reprinted by Hugo Gernsback. Before moving on to the republication and repurposing of Luther Trant, I briefly cover the intervening decades in which the founding principle of lie detection and various lie detectors emerged, applied psychology finally had its day in court, and the lie detector found purchase in the public consciousness.

## Unauthorized Science: Mechanical Lie Detection Goes to Court, 1923

During the two decades following the initial publication and collection of the Luther Trant series, the development of lie detection tests began to mimic the predictions of Münsterberg, Balmer, and MacHarg. On the one hand, lie detection began the process of becoming a science called polygraphy, complete with dedicated experimenters, research funding, and laboratory space. On the other hand, lie detection continued to meet much the same resistance to its status and extradisciplinary applicability as did applied psychology. By 1923, at least one form of lie detection, William Marston's lie detection test using a sphygmomanometer, was deemed inadmissible in American courts. In this section, I detail the rise and eventually damaged reputation of lie detection in American courts between 1915 and 1923; in the final section, I explain the ways in which lie detection rallied from its early demise to rise again in the public's opinion.

From the machines and techniques I described in the first section, three schools of lie detection emerged, thanks to the efforts of several men working in disparate academic disciplines, including psychology, law, and law enforcement, led respectively by William Marston, John Larson, and Leonarde Keeler.[36] William Marston was the first to publish an academic article on the connections between systolic blood pressure and deception in the 1910s.[37] This paper, entitled "Systolic Blood Pressure Symptoms of Deception" (1917), was originally part of his dissertation that earned him a doctorate in psychology from Harvard University. The latter two men, Larson and Keeler, were both protégés of August Vollmer;[38] they spent the greater part of their careers at war with each other over the development, distribution, potency, and proper protocols associated with lie detection machines in their various forms (Alder 2007).[39]

Much like the fictional Luther Trant, Marston, Larson, and Keeler represented the young blood of experimental psychology, law, and police work, respectively: students able to adopt their mentors' work and better adapt it to practical and public applications. So, while Münsterberg had already experimented with lie detection, Marston further advanced the technique by introducing it to judicial court cases and cases of domestic unhappiness. After his initial publications and presentations (between 1913 and 1922), the systolic blood pressure test for deception was taken up and modified by several others who became the core of a new discipline, polygraphy. By 1921, John Larson was experimenting with lie detection under the guidance of August Vollmer; in 1926, Leonarde Keeler (a student of Larson) invented the portable polygraph instrument along with several tests that allowed for the normalization of readings.[40]

Often working in conjunction with each other (as cospecialists or as teachers and students), writing forewords to each other's books and referencing each other's work, these men represented—at least in the early years—what Bruno Latour would characterize as a "network of allies."[41] Marston, Vollmer, Larson, Keeler, and later Fred Inbau did not work alone; instead they relied on mutual authorization. In his 1932 introduction to John Larson's *Lying and Its Detection,* for example, Vollmer lauds the experimental work of his fellow scientists, noting that "Dr. Larson and other scientific workers, like Marston, are blazing a trail that must ultimately lead to fertile fields. Every encouragement and aid should be given to these tireless pioneers" (x).

Although many proponents of lie detection desired to engage disciplines outside of law enforcement and to "stimulate" (Inbau 1942, v) interest in the progressive successes of science's newest machine, the history of mechanical lie detection is fraught with questions of legitimacy and inclusion. Internal strife often divided the polygraph pioneers. Of central concern were disagreements about the effectiveness of the technology itself and arguments about purist versus populist science. Though many scientific and criminological reports of the 1920s and 1930s claim accuracies of 95 to 100 percent with the "truth machine," Fred Inbau, Christian Ruckmick, and Leonarde Keeler admit to more modest positive results of around 70 to 80 percent (Ruckmick 1938; Keeler 1934, 1930; Inbau 1934, 1935a, 1935b, 1935c). When Keeler was asked about the singular strength of the lie detector's results in a 1935 courtroom hearing, he admitted, "I wouldn't want to convict a man on the grounds of the records alone" (Vollmer 1937, 134).

Aside from evidential debates, personal disillusionment and internal strife not only challenged but often delegitimated the machine and its proponents. Some polygraphers, such as John Larson, later recanted their initial belief and involvement in the lie detector's development. In a 1961 article concerning the analysis of lie detector evidence, Larson admitted, "I originally hoped that instrumental lie detection would become a legitimate part of professional police science. It is little more than a racket. The lie detector, as used in many places is nothing more than a psychological third-degree aimed at extorting confessions as the old physical beatings were. At times I'm sorry I ever had any part in its development" (Lykken 1998, 29). Others, such as Fred Inbau, challenged the results and methods of fellow polygraphists even during the 1920s and 1930s. In his review of Marston's book *The Lie Detector Test,* Inbau scoffs at Marston's "exaggerations," argues that this "book is practically useless," and remonstrates with Marston for claiming to be the sole creator and originator of the lie detector (1938, 307).[42] As a result of such diatribes, internal cohesion between polygraphists was ultimately neither sound nor durable, and such divisions often made external "allies" wary of endorsing what they viewed to be a problematic technology.

Of particular interest is the judicial reaction to the polygraph, because it highlights the necessity of accruing allies in a search for authorization while it simultaneously reveals the uncertain internal dynamics of applied psychology and, by the 1920s and 1930s, police science. In 1923, a young black man named James Alphonzo Frye was to be tried for the second-degree murder of Dr. R. E. Brown. Frye initially confessed—after days of grueling interrogation—but later took back his admission. William Marston was called in to administer his systolic blood pressure test for deception and potentially testify in court. Despite the defense's petition, Judge McCoy denied the admissibility of Marston and his test. When the case came before the appellate court later that same year, Justice Van Orsdel not only upheld McCoy's decision but established a precedent that influenced the admissibility of scientific evidence for over seventy years.[43] In his written opinion, Van Orsdel had the following to say.

> Just when a scientific principle or discovery crosses the line between the experimental and demonstrable stages is difficult to define. Somewhere in this twilight zone the evidential force of this principle must be recognized, and while courts will go a long way in admitting expert testimony deduced from a well-recognized scientific principle or discovery, the thing from which the deduction is made must be sufficiently established to have gained general acceptance in the particular field in which it belongs. (*Frye v. U.S.*)

The decision draws attention to a fissure in polygraphy's history—a rupture in the foundations of scientific fact production and attempts to discipline knowledge, assigning authority to "the particular field in which [a technology] belongs." In particular, Judge Van Orsdel's reasoning reveals the desire for consensus and the belief that science must complete its deliberations before emerging into the public or judicial sphere. By calling attention to disciplinary boundaries and their potential for authorization, Judge Van Orsdel's decision highlights the troubled interdisciplinarity of lie detection and one of its parent disciplines, applied psychology. Yet, instead of isolating psychologists, lawyers, and police-men within their respective fields, *Frye v. U.S.* helped usher in another era of cross-disciplinary legitimation for the lie detector in both fiction and science.

### Strategic Reemergence: Hugo Gernsback and Luther Trant, 1925–30

Far from losing faith in lie detection after the *Frye* case, proponents of lie detection constructed a network of science and fiction in which they—and their machines, techniques, and theories—were indispensable to the public. Marked by a confluence of sensationalism and edutainment, scientific texts, and popular nonfiction of the era worked to enliven themselves through true crime stories, while literary authors and editors of the period sought to scientize their narratives by referencing and us-ing the very technologies they sought to validate as scientific. In 1925, Hugo Münsterberg's *On the Witness Stand* was reissued. His directive to bring this discussion to the "wider tribunal of the general reader" sounded all the more appropriate after the *Frye* case conclusively proved that "the lawyer and the judge and the juryman are sure that they do not need the experimental psychologist" (1908, xi). Indeed the law versus psychology rhetoric emerged anew after the *Frye* case.[44]

In the scientific and popular nonfiction literature, Münsterberg's reissued call for a "wider tribunal" took many visible and hybrid forms; most referenced the fiction of Luther Trant's major predecessor, Sher-lock Holmes. In 1930, editors of the *American Journal of Police Science* translated and republished Edmund Locard's work on "The Analysis of Dust Traces." Locard, founder of the first modern crime laboratory in France in 1910, argues that "the police expert, or an examining magis-trate, would not find it a waste of his time to read Doyle's novels . . . and one might profitably reread from this point of view the stories entitled *A Study in Scarlet, The Five Orange Pips,* and *The Sign of Four*" (1930, 277). Henry Morton Robinson's *Science Catches the Criminal* (1935), a veritable encyclopedia of criminalistics from their inception through the mid-1930s, begins by citing Sherlock Holmes and goes on to combine histor-ical facts with sensationalized true crime stories. In fact, a 1935 *New York Times* book review praises Robinson's work for this very reason, lauding the virtues of the "dramatic" stories included throughout and noting that "his book deserves the widest popular attention, for its theme [of technologies for criminal investigation] is one of universal and fundamental importance upon which there needs to be a very general spreading of enlightenment. And, besides, it is better stocked with thrills than a detective story" (Kelly 1935, BR4). As late as 1941, T. G. Cooke's *The Blue Book of Crime,* an advertisement for and explanation of current forensic techniques in fingerprinting, argues that "the mystic days of the supersleuth may be gone, but romance and adventure still live in this profession—excitement still thrills—for the trained man of today finds his work as varied, as stimulating as ever did a Sherlock Holmes—his discoveries as animated and stirring" (1941, 5). Finally, as detailed later in this section, Hugo Gernsback repurposed and republished several Luther Trant stories concerning lie detection between 1926 and 1930. His choice of Luther Trant is particularly notable given the prominence of Sherlock Holmes in the media.

Gernsback, who published several science and technology magazines during the early decades of the twentieth century,[45] was, by the 1920s, in the business of promoting a new genre of "scientifiction." To advance his new genre, Gernsback republished and thereby reclaimed various short stories as scientifiction in what he termed "A New Sort of Magazine,"[46] *Amazing Stories,* initially published in 1926.[47] The so-called scientific romances of Edgar Allen Poe, Jules Verne, and H. G. Wells were some of the first tales to be redubbed "scientifiction," "a charming romance intermingled with scientific fact and prophetic vision" (Gernsback 1926, 1).

By the 1930s Gernsback would begin to codify scientific detective fiction in his essay "How to Write 'Science' Stories" (1930a). This piece should not be confused with Gernsback's other submission suggestions for "scientifiction" or science fiction, generally, as it pertains most particularly to the application of science to problems of crime and law. Take, for example, his first pair of "do's" concerning the subgenre that he hereby defines.

> (1) A Scientific Detective Story is one in which the method of crime is solved, or the criminal traced, by the aid of scientific apparatus or with the help of scientific knowledge possessed by the detective or his coworkers. . . . (2) A crime so ingenious, that it requires scientific methods to solve it, usually is committed with scientific aid and in a scientific manner. (27–28)

As with scientifiction, Gernsback argued that scientific detective fiction would become preeminent. "We prophesy that Scientific Detective fiction will supersede all other types. In fact, the ordinary gangster and

detective story will be relegated into the background in a very few years.
. . . Literary history is now in the making, and the pioneers in this field
will reap large rewards" (1930a, 28). While his hopefulness about both
scientifiction and scientific detective fiction is certainly biased and self-
interested, Gernsback's definitions and the popularity of his magazines
ultimately ushered in a new genre that did stand the test of time: science
fiction.

Part of his success can no doubt be attributed to the fact that in all of
his publications, Gernsback was invested in an "edutainment" model
that valued stories that could teach his readers something about science,
technology, and progress while also entertaining them. In his editorial
introduction to *Amazing Stories* Gernsback extols the virtues of his magazine by noting, "Not only do these amazing tales make tremendously interesting reading—they are also always instructive. They supply knowledge that we might not otherwise obtain—and they supply it in a very
palatable form. For the best of these modern writers of scientifiction
have the knack of imparting knowledge, and even inspiration, without
once making us aware that we are being taught" (1926, 1). His vision,
like Münsterberg's, is centered around both guiding and relying upon
the power of a lay audience's support.

In 1926 and 1927, Gernsback republished not one but four Luther
Trant stories in *Amazing Stories:* "The Man Higher Up," "The Eleventh
Hour," "The Hammering Man," and "The Man in the Room." He would
later republish them for a second time in *Scientific Detective Monthly* between 1929 and 1930, with the addition of "The Fast Watch."[48] The
sheer volume of space devoted to these Luther Trant stories and the fact
that they were published alongside the likes of Edgar Allen Poe, Jules
Verne, and H. G. Wells indicates Gernsback's estimation of them. The
Luther Trant stories were arguably also selected because they fit Gernsback's model of edutainment: they educated audiences through exposi-
tory lumps and embedded textual representations of lie detection technologies, they worked from and for a prophecy model forecasting legal
validation, and they reassured the public about the possibility for crime
prevention and criminal punishment via technology in an era of increasing police corruption and gangster violence.

Gernsback's purpose can be seen in the explanatory text boxes he in-
serts into the stories. In the case of Luther Trant, Gernsback includes information about the technologies used by the psychological detective, in
order to highlight what is new and different about Trant's stories: as

mentioned earlier, they represent a new school of scientific detective fiction popularized in America and distinct from its earlier American and European counterparts in that they feature instrumentation.

If *Amazing Stories* was intended to inform and excite a lay audience about nascent technologies, one of Gernsback's later magazines, *Scientific Detective Monthly*,[49] was designed to educate the public about how these technologies have been put to proper use by the police. In an early editorial, "Science vs. Crime" (Jan. 1930), Gernsback writes, "I sincerely believe that *Scientific Detective Monthly* will not only prove to be a creative force in this type of literature, but *actually help our police authorities in their work*, by disseminating important knowledge to the public, and be also a constant warning to the criminal that, with adequate scientific laboratories, crime will have less and less chance to survive undetected" (84; emphasis added). The latter part of this statement is an indirect speech act: the likelihood that criminals will read, be educated, and be reformed by *Scientific Detective Monthly* is slim at best; however, Gernsback is speaking not to the criminal but to those who could fund and support the "scientific laboratories," the existence of which will surely produce new and improved instrumentation and techniques for crime fighting.

Gernsback saw himself as a disseminator and mediator of not literature but scientific knowledge. As in Balmer and MacHarg's editorial, Gernsback attempts to authorize his magazine as dealing in fact, not fiction: "While *Scientific Detective Monthly* may print detective stories whose scenes lie in the future, it should be noted that whatever will be published will be good science. We describe no fictional apparatus, no methods not based upon present-day science" (1930a, 84). When he does present an apparatus whose acceptance and deployment are marginal, as in the case of lie detection instruments, Gernsback makes predictions for their eventual authorization in offset text boxes, predicting, as he does, for example, in Balmer and MacHarg's "The Fast Watch" (1930b), that "tests of this nature will be in actual use at a not too distant future to allow the criminals to reveal their own guilt or to establish their innocence."

In the spirit of Hugo Münsterberg and other popularizers, Gernsback's choice of Trant stories in both *Amazing Stories* and *Scientific Detective Monthly* is specifically invested in not only educating the public but illustrating the adoption of lie detection technologies by law enforcement detectives and the court system. Indeed, the stories selected by Gernsback do not represent the breadth of Luther Trant's adventures cataloged in

*The Achievements of Luther Trant* (1910). Instead, they are a selected set of narratives that explicitly demonstrate the chronoscope and four different technologies that can be used for lie detection (the pneumograph, plethysmograph, galvanometer, and sphygmomanometer), which were later combined into one machine known as the polygraph.

Aside from introducing the public to the instruments, Gernsback was invested in a visionary model of science fiction, for which he would assume the role of disseminator. "Many great science stories destined to be of an historical interest are still to be written," Gernsback argued, "and *Amazing Stories* magazine will be the medium through which such stories will come to you. Posterity will point to them as having blazed a new trail, not only in literature and fiction, but in progress as well" (1926, 1). In the case of the Luther Trant stories, Gernsback foregrounds his role in bringing this technology to the public for approval, even after it has been deemed inadmissible by the courts. In an offset text box embedded in the first page of "The Man Higher Up," Gernsback notes that "while the results of psychic evidence have not as yet been accepted by our courts, there is no doubt that at a not distant date such evidence will be given due importance in the conviction of our criminals" (Balmer and MacHarg 1926, text box, 793).[50] In the third story in this miniseries, "The Hammering Man," another text box mimics Balmer and MacHarg's own fiction-science paradox, noting that "the strange part about it all is that although the story is written as fiction, the results can be obtained readily any time today, as the instruments used are well known and can be found in any university and up-to-date college laboratory" (Balmer and MacHarg 1927b, text box, 1118). Moreover, the republication of these four stories not once, but twice—first as scientifiction/science fiction and then as scientific detective fiction—mirrors their eventual, though fraught transition from speculative technology to applied police science.

Lest lie detection appear dangerous and invasive, the stories reprinted by Gernsback also inform the public about the numerous ways that such technologies promise to protect and intervene on their behalf. It is no coincidence that Gernsback's reprints emerged in the social climate of the late 1920s and early 1930s during which political corruption often ruled the police force and organized crime ruled prominent cities like New York and Chicago (Powers 1983; Walker 1977, 1998). Lie detection, along with the move toward police professionalization, became more widely and readily acceptable because they promised to sanitize, organize, and control the objectionable behavior of law enforcement officials. However, instruments like the lie detector, which threatened to reveal inner truths and objectify individuals, were frightening manifestations of technological power, especially in the hands of already questionable law enforcement agencies. Balmer, MacHarg, and (through his editorializing) Gernsback may have been responding to this social unease when they (re)packaged lie detection as a force of progressive social change: to mitigate police corruption, level class distinctions, and erase the advantage stereotypically attributed to populations of various races.

The last story reprinted by Gernsback in *Amazing Stories,* "The Man in the Room," was the first story published by Balmer and MacHarg. As we have already seen, it directly confronts police corruption as Trant dialogues with his mentor, Dr. Reiland, about the horrors of third-degree interrogations and the inefficient legal system. By reprinting this particular story, Gernsback echoes Münsterberg's sentiment from *On the Witness Stand* that "the vulgar ordeals of the 'third degree' in every form belong to the Middle Ages, and much of the wrangling of attorneys about technicalities in admitting the 'evidence' appears to not a few somewhat out of date, too: the methods of experimental psychology are working in the spirit of the twentieth century" (1908, 109). Both characterizations of the third degree and experimental psychology rely on a narrative of progress that values psychology as a forward-looking discipline capable of transforming other fields as well as society at large. I further explore the pattern and implications of this progress narrative in chapters 3, 4, and 5; importantly, the narrative of progress heard here will be repeated at key moments in the cultural history of lie detection explored throughout this book.

In addition to rendering the third degree obsolete, the lie detection instruments—and narratives—reprinted by Gernsback also promise to level class distinctions. By the end of "The Man Higher Up" it becomes clear that the president of a corporation and the common thief are equally susceptible to the instruments of experimental psychology. Neither can control the physical changes that incriminate them. After administering the test, Trant declares to the president, Welter, "you betrayed yourself uncontrollably, unmistakably" (Balmer and MacHarg 1926, 801) and remarks "it's some advance isn't it, Rentland, not to have to try such poor devils alone [the checker and the dock superintendent]; but, at last, to capture the man who makes the millions and pays them the pennies—the man higher up?" (1930d, 867). Years later, the popularity of these instruments rested in part on their marked ability to destroy the myth of untouchability associated with the upper classes. Not only is the president of the company charged, but "modern practical psychology" is given credit "for proving the knowledge of the man higher up" (1926, 796), thereby making him vulnerable to psychological investigation and examination.

Finally, "The Eleventh Hour" directly addresses the leveling of racialized propensities for evading detection, which we glimpsed in the *Frye v. U.S.* case and will confront again in contemporary brain-based detection. In this particular narrative, Trant is confronted with a death that occurs under very abnormal circumstances: five shots fired, only four casings found, the abused wife is suspected, strange murmurings are heard in the night, strange shoeprints are found in the snow. When the evidence points to murder and a "Chinaman" as suspect, the rank-and-file police nearly concede the case: "if it was a Chinaman you'll never get the truth out of him" (Balmer and MacHarg 1927a, 1049). Initially Trant agrees: "I know . . . that it is absolutely hopeless to expect a confession from a Chinaman; they are so accustomed to control the obvious signs of fear, guilt, the slightest trace or hint of emotion, even under the most rigid examination, that it had come to be regarded as a characteristic of the race" (1049). However, this crestfallen moment is then transmuted into an opportunity to showcase "the new psychology [that] does not deal with those obvious signs; it deals with the involuntary reactions in the blood and glands which are common to all men alike—even to Chinamen!" (1049). Several "Chinamen" are brought in, attached to the galvanometer and tested for veracity; Sin Chung Min is found to be the perpetrator of the crime, and an analysis of his exam—particularly at those moments when he is queried about his accomplices—implicates the other three men.

The progress narratives and social leveling highlighted in the selective reprinting of the Luther Trant stories helped to characterize lie detection as a dynamic corrective to uncivilized practices and unfair advantages (be they economic privilege or—in a twisted sense—racialized talents for evasion). Lie detection paradoxically promises to enhance systems of power while remaining a tool of social justice: lie detection is the epitome of Münsterberg's mental microscope, providing access to those hidden structures, beliefs, and lies that have the power to undermine the functioning of society. Gernsback's argument jibes with the sentiments of Balmer, MacHarg, and Münsterberg: judges, lawyers, and police need to recognize the power and potential of applied psychological instruments. Even if courts will not admit testimony from the devices, lawyers scoff at the idea of objective testimony, and police dislike being displaced by machines, the validation of lie detection can be taken to "the wider tribunal of the general reader" in one form or another. As explained in "The Eleventh Hour," instrumental testimony is confession enough for the public and even for suspects. After his session with the galvanometer, the "Chinaman" commits suicide in his cell. "He considered what we learned from him here confession enough," Trant explains. "You can safely consider your case settled" (Balmer and MacHarg 1927a, 1051).

In total, Gernsback's republished and repurposed science fiction made lie detection visible, not as a technology that had failed in the courts, but as a technology ripe with potential—one that could not and should not be restrained by conservative detractors, including judges, lawyers, and police officers. His use of Luther Trant in particular gave voice to a crossover scientific detective capable of applying psychological techniques and instruments to the problems of criminal investigation and criminal law.

Although his "achievements" were published three times over in popular magazines between 1909 and 1930, Luther Trant is an unsung and seemingly unlikely hero in the history of lie detection. His adventures combine technology, law, psychology, and criminal investigation in novel and prophetic ways; his character represents the consummate consulting psychologist; his enthusiasm and knowledge promise to educate the public, eradicate crime, and bring a novel approach to detecting criminality. But Trant would not have made such a splash in the early decades of the twentieth century without the help of his authors and editors who were attuned to debates about the application of psychology to law and detective work championed by Hugo Münsterberg.

The interstices between Luther Trant's adventures, Münsterberg's dreams, and the courtroom represent a moment when fiction helped legitimate an applied science and its instrumentation that could not find authorization in courts of law. If scholarship on law and psychology has focused largely on the controversies surrounding Hugo Münsterberg (Blumenthal 2002; Hale 1980) without examining the results of his call for a wider tribunal in the fiction of the first and second decades of the twentieth century, then the wider history I trace in this chapter, framed as it is by both Münsterberg's collection and *The Achievements of Luther*

*Trant,* helps us to better contextualize the arenas in which Münsterberg's theories and aspirations held sway. Although he and his disciples were ultimately unsuccessful in persuading the lawyer and the judge, they were capable of creating a public following for lie detection via literature, American scientific detective fiction in particular.

As Hugo Münsterberg assumed, "The poets know it well"—"it" being the basic principle behind lie detection: that emotions will be revealed through the body. I will return to fiction in later chapters; but in the next one I analyze the experimental work of William Marston, perhaps the most (in)famous of Münsterberg's students. As we shall see, his "mock crimes" along with his qualitative and quantitative evaluations of his subjects bring us back to the gray area between science and the imagination, truth and lies.

# The Science of Lying in a Laboratory

## William Marston's Deceptive Consciousness, 1913–22

> The difference between telling stories and acting realities
> isn't so large.
> —JOHN LAW AND VICKY SINGLETON (2000, 769)

Known for analyzing "baby parties" at the local sorority, creating the iconic, lasso-toting Wonder Woman, and penning the salacious history of Julius Caesar's private life,[1] William Moulton Marston also invented several consequential lie detection protocols in the early decades of the twentieth century. A Harvard-trained lawyer (1918) with a PhD in psychology from the same institution (1921), Marston was convinced, like his mentor Hugo Münsterberg before him, that the emotional disturbances he associated with lying would necessarily produce physiological reactions. Like his fictional counterpart Luther Trant, Marston represented a revitalizing force in psychology: he knew, for example, that psychophysiological research was taking a new direction, away from the galvanometer and word-association tests of Francis Galton, Carl Jung, and Sigmund Freud and toward measures of respiration and blood pressure.[2]

In Marston's dissertation and first academic article on lie detection,[3] his goal was to describe, illustrate, and quantify what he termed "the deceptive consciousness" (1917, 153). According to Marston, the "deceptive consciousness" was associated with emotions of rage and fear, accessible through changes in physiology, and evocable in a laboratory.

Marston was not interested in the "objective" truth or falsity of a subject's narrative but in the subject's own feelings, knowledge, or perception of that narrative. In his conclusions, for example, Marston remarks that "the behavior of the b.p. [blood pressure] does not act as the least indicator of the objective validity of the story told by any witness, but it constitutes a practically infallible test of the *consciousness of an attitude of deception*" (1917, 162; emphasis in original). Marston sought to prove that this deceptive consciousness not only existed but was standardizable: a "'lying complex' sufficiently uniform in different individuals to be experimented upon as a unit" (117).

Not surprisingly, Marston's showmanship in the 1930s and 1940s and his increasingly global vision of the political and personal applications[4] of the lie detector test have eclipsed his scientific innovations in laboratory-based lie detection. In lie detection scholarship, Molly Rhodes, Geoffrey Bunn, and Ken Alder have all focused on Marston's popular appearances and publications, not his series of important academic articles that appeared in print between 1917 and 1927. These articles introduce several features of lie detection experimentation that continue to influence contemporary protocols: first, that there is such a thing as the deceptive consciousness; second, that the deceptive consciousness can be induced by playing one of three roles I have termed the *antagonist,* the *false witness,* and the *mock criminal;*[5] and finally, that there is a modern and acceptable language for interpreting graphic data produced during laboratory studies of deception. Taken together, Marston's experimental protocols have shaped the way we understand and classify acts of prevarication as "working objects" (Daston and Galison 1992, 85) of laboratory science.

In what follows, I first situate Marston's experimental work within the broader context of emotions and laboratory science of the late nineteenth and early twentieth centuries. I then analyze Marston's laboratory research conducted between 1913 and 1922;[6] these studies were the first published records of lie detection experiments in America. Included here is a detailed description of Marston's experimental design and protocols, followed by an analysis of several key assumptions that inform Marston's conceptualization and deployment of the deceptive consciousness.

## Invoking: The Mechanics of Emotion

Developing laboratory culture of the latter half of the nineteenth century was founded on an ideal of objectivity and depended on replicable experiments that could be witnessed by fellow scientists.[7] Introducing emotions into this setting challenged the very foundations of experimentation and threatened the tentative separation of science from its "softer" counterparts in the humanities (Dror 1998). In their affective expression, emotions such as fear, guilt, excitement, and anger demonstrated the capriciousness of the body and were deemed inappropriate for the controlled setting of the modern laboratory. Like the problem of fatigue, explored by Anson Rabinbach (1990), emotions served as a reminder of the body's nonpredictable status.

Yet, during the latter half of the nineteenth century, physiologists and psychologists welcomed emotions into their laboratories.[8] Emotions were rendered appropriate for and acceptable to experimental science in three respects: they were expressed quantitatively; they were separated from (feminized) "affect"; and they implied direct access to the natural body (Dror 1999a, 1999b, 2001b). First, experimenters worked diligently to transform emotions into mechanically recordable, reproducible, and evocable phenomena. Instruments such as the cardiograph, pneumograph, and sphysmograph helped to transform emotions from subjective, individual experiences into objective graphic images—emotions as numbers that could be interpreted by experts (see chapter 1). Indeed, "the number was an important technology for the refraining of 'emotion' and its integration into the discourse of the laboratory" (Dror 2001a, 371). Graphic representations of measurable phenomena such as blood pressure, the electricity of the skin, and the rate of respiration separated the body into controllable and partitioned fragments, which could then be correlated to various, otherwise not directly quantifiable, emotional states.

Once displayed graphically, emotions were no longer secret, hidden, private, and feminized phenomena (Squier 2004). The emotions invoked and measured in laboratory settings were distinguished from the emotions of sonnets and love stories (Dror 2001a) and separated from introspective reports about interior states or feelings. Indeed, experimentally induced emotions were carefully controlled and managed through the sterilizing and objectified act of intervention (Armstrong 1998).

By standardizing the interpretive leap between physiology and emotion, mechanical records come to represent objective knowledge about the body. Emotional inscription technologies in particular invoked ideals of permanence, access, and the potential for duplication. "Borrowed from the vocabulary of archaeology, monuments, statuary and the like, the notion of inscription reinforced the notion of the graphic trace as a storage medium, a material system of memory that would conserve in a permanent form the essential meanings of nature and bring them forth on demand" (Brain 1996, 22). Indeed, one purpose of the graphic record produced in lie detection exams was to create specific categories of criminals that were reified by the empirical evidence of documentation.[9] They were part of a reproducible experiment, a public record, a documented life "that places individuals in a field of surveillance also situates them in a network of writing; [the documented life] engages them in a whole mass of documents that capture and fix them" (Foucault 1995, 189).

The machines that produce these records also become "a means of circumventing the deceptions that have come to be associated with embodied sight" (Cartwright 1992, 140). Indeed, scientists claimed that their machines enabled emotions to write themselves.[10] As feminist critics of science have pointed out, this kind of weak objectivity (Harding 1991) presumes that science is value neutral, that it merely holds up a mirror to nature. And yet, for physiologists and psychologists working in the early twentieth century, this claim of transparency was particularly important precisely because it shrouded the active construction of a new explanatory language. Along with the shift in ways of seeing I began to outline in chapter 1, the language used to describe mechanically produced images "was also a new invention" of experimenters (Dror 1999b, 373). While they were working with familiar bodily signs, their representations and interpretations were radically new. Deception, for example, was said to be a conglomerate of several simpler emotions, including fear and anger. To draw such conclusions, experimenters, such as William Marston, had to simultaneously craft an ethos for themselves and their instruments. The process required innovation because,

despite [experiment]ers'] assertions to the contrary, and their claims for continuity between their own language of emotions and familiar emotional phenomena . . . there were no practitioners from whom they could learn how to interpret emotions from their graphs; no guidelines to tell them what actions to take in order to eliminate alternative interpretations; no explicit conventions when it came to representing emotions in graphic and numeric language; and no canon of existing emotional texts. (Dror 1999b, 373)

Marston himself admits this lack of context in the introductory remarks to his dissertation, noting that "at the time this research was begun in 1913 there could not be said to be any historical background whatever in the field of psychology of deception. There was some literature upon the association reaction time diagnostik and this work, while none of it bears squarely upon the point of deception itself, nevertheless, furnishes the only real historical approach" (1921b, 3).[11]

The latitude afforded by this process allowed experimenters to create a set of new "working objects," which can be defined as "any manageable, communal representatives of the sector of nature under investigation" (Daston and Galison 1992, 85). Like inscription technologies, working objects alleviate the problem of embodied experience: a working object is neither sensorial data nor the potentially subjective end product of science (a theorem or concept). They are, as Lorraine Daston and Peter Galison argue, "the materials from which concepts are formed and to which they are applied" (85). William Marston's working object was "a significant lying curve" (Marston 1917, 156) from which he extrapolated his concept of the deceptive consciousness and to which he applied his quantitative and qualitative interpretations.

## From Working Object to Major Concept: Defining the Deceptive Consciousness

William Marston began his experimental career in the Harvard Psychological Laboratory—one of the first, and most cutting-edge, experimental psychology laboratories in America (Landy 1992, 788–98; Littlefield 2010). As we saw in chapter 1, the laboratory was under the direction of Hugo Münsterberg who emigrated from Germany at the request of William James. Münsterberg, himself a protégé of Wilhelm Wundt, brought his own vision for psychology with him across the ocean: the young German scientist was interested in applied psychology. Underlying nearly all of Münsterberg's theories was the idea that psychological states and emotions manifest themselves in and through the body.

Although Münsterberg died just as Marston began his advanced degree work, his pupil proved an apt disciple. When he developed the applied systolic blood pressure test for symptoms of the deceptive consciousness, Marston made manifest Münsterberg's vision for applied psychology in theory and in practice. In theory, Marston seconded Münsterberg's ideal of a mental microscope that would provide access to "the drama which goes on inside" (Münsterberg 1908, 114). As Marston would go on to describe them, the symptoms of the deceptive consciousness involved a measurable struggle between conscious, voluntary suppression and involuntary, unconscious expression. These symptoms were best encapsulated in the working object of "a significant lying curve," which "is a function of the struggle between the involuntary impulse to express fear in response to awareness of danger, and the voluntary focusing of attention to exclude the fear from consciousness" (1917, 156). To call this process a "struggle" between physiological reactions and psychological strategies establishes body and mind as warring substrates; their battle can then be recorded by monitoring changes in the autonomic nervous system—via the systolic blood pressure—and correlating these changes with self-reported data about states of consciousness.

Practically, the test consisted of Marston intermittently taking blood pressure readings using a Tycos sphygmomanometer[12] as the subject answered yes/no questions; Marston recorded his findings in a table and charted the results in graphic form.[13] In general, "the systolic blood pressure was found to rise between 10 and 50 millimeters during deception and this rise was found to follow a very constant and regular form of curve with the climax of the curve being closely correlated with the subjective crisis or climax in the subject's deceptive story as introspectively reported" (1921b, 4).[14] One of the keys to Marston's experiments evident in this passage is his mixture of qualitative (graphic) and quantitative (introspective) data collection. The former figures the body as a mechanistic organism whose physiological performance can be quantified, codified, and compared.[15] The latter figures the body as a subject capable of understanding and voluntarily divulging affect. "Only [through a holistic analysis]," Marston argues, "can the curiously close correlations of introspection and pressure record, the individual peculiarities, and certain interesting mixtures of truth and lying be considered" (1917, 131). I will return to the "interesting mixtures of truth and lying" later; here, allow me to focus on Marston's holistic analysis of internal drama.

To display a deceptive consciousness under the laboratory conditions I describe in a moment, subjects must not only deviate from "the truth"—which is clearly defined within the contours of the experiment—but must also be mindful of and define their deviations as deception. A certain tautology is at work here and built into the fabric of Marston's experimental design and results. If a subject did not confess deception to themselves,

then their systolic blood pressure record would not reveal any changes indicative of the deceptive consciousness; and if the graphic record did not
reveal a rise in systolic blood pressure for a given question, then the subject must not have felt an attitude of deception.

Marston chose to measure systolic blood pressure specifically[16] because it afforded a mechanized image of the "struggle" he sought to isolate. Following the work of his contemporaries W. B. Cannon and Alfred
Binet, Marston argues that the systolic blood pressure is less likely to be
influenced by pain, intellectual activity, and minor affective states; more
important, systolic blood pressure is suspected to be harder to consciously control or inhibit.

> First, the use of the systolic eliminates the local effects of minor affec-
> tive states; secondly, it eliminates the important and irrelevant factor
> of intellectual work; thirdly, it is less susceptible to modification by
> physical pain than is the diastolic; and fourthly, it tends to record only
> the unequivocal changes in the b.p. system brought about through in-
> crease of heart-beat unimpeded by inhibitory reflexes or antagonistic
> functioning of the vaso-motor apparatus. (1917, 121–22)

Viewed as a machine capable of divulging symptoms of deception, the
body does not betray the mind so much as the physiological responses
that are *not* associated with pain, intellectual work, minor affective states,
and the "heart-beat unimpeded by inhibitory reflexes." The deceptive
consciousness depends on a subject's cognizance, yet it is best measured
via the autonomic nervous system: the one physiological system that is
not under conscious control. We will encounter similar logic in the design and application of contemporary lie detection technologies including EEG-based Brain Fingerprinting and fMRI in chapter 5. These contemporary technologies seek to isolate other unconscious processes: the
electrical activity or hemodynamics of the brain.

Detecting the other half of the struggle, evident in Marston's performance analyses, involves several layers. First, he asked his subjects to in-
trospect their feelings about their own performance. Subjects introspected that they felt "disgust," "shame," "nervous," "embarrassed," "like
during an exam," "tense," "sneaky," and "excited," among other emotions. In the systolic blood pressure experiment, one subject even introspected that he experienced "feeling like 'stage *fright*'" (1917, 141).[17]
Second, Marston categorized the nature and plausibility of each subject's narrative. In a section titled "Individual Results," Marston explored

each subject's systolic blood pressure record, introspections, truthful or deceptive narratives, and demeanor. While it is difficult to determine what would be a "good" or, at the very least, "acceptable" narrative, Marston's analysis makes one thing very clear: his subjects—like actors and comedians—were judged on the details, plausibility, consistency, delivery, and physical demeanor of their performance. "E's stories were racy, dramatic, but inaccurate and careless" (141), for example; while "J's stories were plausible, consistent, but not ample; and were told in a straightforword manner" (151). In some cases, subjects failed to construct good lies but were still commended for their ability to confuse a mock jury. "The stories composed by Subject A were," for example, "on the whole, very poor alibis. They were rambling, indefinite, rather wild, and very improbable, yet while telling the truth this subject managed to convey, by his peculiar manner of narration, the impression that he was lying, so that the jury found it very hard to judge correctly in any case" (131). In short, Marston's judgments of success or failure are contingent; they depend on far more than the enumeration of emotion through the systolic blood pressure record.

Marston's performance analysis highlights the mutual imbrication of mechanical and human interpretation. And yet, in his experimental protocols and analyses, Marston artificially separates the two: he plays the examiner who records and interprets the systolic blood pressure, while a human jury attempts to judge deception solely via observation of the false witness or mock criminal. This formation allows Marston to establish his mechanical and interpretive accuracy as compared to that of a jury. Not surprisingly, he nearly always defines himself as expertly capable and finds the jury lacking. "The experimenter, basing his judgment entirely upon the b.p. behavior, made 103 correct judgments and 4 erroneous ones. . . . The b.p. judgments were, then, 96 per cent. correct" (1917, 159). He is less than flattering of the jurors, noting "that the jurors, not the subjects, are the ones to be divided into successful and unsuccessful classes" (160). The jurors had success rates of 10 to 85 percent, with the average rate of the twenty jurors being 48.25 percent.

In order to market his ideas, Marston's rhetorical approach is reminiscent of both Hugo Münsterberg, who believed in applying psychology to social ills, including crime, and the fictional Luther Trant, who desired to have his psychological techniques adopted by criminologists, lawyers, and police alike. Marston's earliest experiments were published cross-disciplinarily both in psychological journals, including the first volumes of the *Journal of Experimental Psychology*, and criminological journals.[18] And, long before the *Frye* case, Marston was sensitive to the requirements for juridical admissibility of applied lie detection. In a 1920 publication, he noted that "our present trial systems demand, not only expert testimony, but the qualification of the expert by proof of the existence of a commonly known and recognized body of scientific fact upon which the expert bases his opinion" (73). By 1921, he was optimistic "that long before the legal problem of such tests is solved the fundamental psycho-physiological elements will be rather clearly analyzed out" (1921a, 570).

To make his case compelling outside the laboratory, Marston extrapolated from his working object (the significant lying curve) to his major concept, "the deceptive consciousness." Marston hypothesized that the deceptive consciousness is uniform among individuals and that "we must, therefore, seek objective, quantitative measurements of the psycho-physiological symptoms of the deceptive consciousness" (1920, 73). Marston's reasoning comes again, in part, from a lengthy analysis of Walter Cannon's work on emotion and physiology[19] and Vittorio Benussi's work on deception and respiration. Building on Benussi's research, Marston argued that "Benussi's results, indicating as they do great definiteness of lying symptoms, are sufficient to warrant the assumption of the uniformity of the deceptive consciousness as a working hypothesis" (1917, 118). Although he later modified this uniformity hypothesis to include two types of deceptive consciousness,[20] he consistently argued that the body will always betray symptoms of deception so long as an attitude of deception is present.

By 1925, Marston defined the deceptive consciousness without reference to either the significant lying curve or the systolic blood pressure; it became simply "a certain definite state of the organism characteristic of a voluntary attempt to deceive another person" (1925, 244). In this process of black-boxing (Callon and Latour 1981, 285), Marston's deceptive consciousness sloughs off its production to become a distinctly measurable phenomenon. Instead of seeking a significant lying curve and extrapolating a deceptive consciousness, the deceptive consciousness is expected to produce a significant lying curve whenever a subject attempts to manage nervous reactions he[21] should not be able to control. Recursively constituted, the deceptive consciousness appears uniform among individuals and is, ostensibly, infinitely evocable within the confines of a laboratory.

## Protocols for Invoking the Deceptive Consciousness

Given his definition of the deceptive consciousness, Marston did not make claims to decontextualized measures of truth or deception but situated deception firmly within a particular complex of emotions, in a given subject, at a particular time. He seemed cognizant of the fact that his hypothesis about the deceptive consciousness could not record all lies, variations, or augmentations. Indeed, he was invested in laboratory protocol that isolated particular kinds of deception. Between 1913 and the 1921–22 academic year, Marston primarily performed the research for five studies related to the systolic blood pressure deception test at Harvard's Psychological Laboratory. During this same period of time, Marston's research found practical applications outside of the laboratory: he performed a study at Georgia's Camp Greenleaf in 1918, which was sponsored by the military, and proposed to evaluate the possibility of training future operatives; his work was also sponsored by the Psychological Committee of the National Research Council,[22] through which he was able to test his techniques on actual criminal defendants in 1917. By 1923, Marston felt confident enough in his technique to bring it to court for *Frye v. U.S.* (see chapter 1).[23]

In his efforts to produce a significant lying curve and thereby uncover the symptoms of the deceptive consciousness, Marston introduced several types of lying into the laboratory.[24] In each experiment,[25] subjects were asked to play one of several roles, which I term the *antagonist,* the *false witness,* or the *mock criminal.*[26] Subjects were drawn mostly from the psychology department's undergraduate and graduate students; as such they were considered to be "trained subjects" (Marston 1920, 74; 1923, 394). These trained subjects were asked to deceive experimenters in several predetermined ways: the *antagonist* disobeyed experimental orders, the *false witness* constructed alibis, and the *mock criminal* committed a staged crime and then lied about her activities. Each subject had his systolic blood pressure monitored at various times: antagonists were monitored as they provided word associations, false witnesses were under surveillance as they constructed their alibis, and mock criminals submitted to the systolic blood pressure exam as they testified about their previously performed activities. A few examples will help illustrate how each role is enacted in the laboratory.

In Marston's first experiment (1913–14), which followed typical word-association protocols established by Galton, Jung, and Freud, subjects were presented with cards on which were printed two columns of words. Subjects were expected to provide associations for each word in a particular order, from top to bottom and left to right, for example.[27] They were also expected to play the role of antagonist at least some of the time. Antagonists were ironically *instructed* to "reverse instructions" (Marston 1920, 75); a reversal meant interacting with the words in the wrong order (bottom to top and right to left, for example). In this type of experiment, deception was defined in terms of information (non)disclosure: "The subject was to deceive the experimenter as to the identity of the cards where instructions were reversed by proceeding to make his associations with the list on these cards as rapidly as on the cards where he obeyed instructions" (75). Marston timed the reactions to each word association in order to determine if the subject was following protocol or reversing instructions. If she was compliant or "truthful," her reaction times were not affected; if she was refusing orders, playing the antagonist, and therefore deceiving experimenters, her reaction times were affected.

What is at issue here are the triadic conflations of truth, correctness, and compliance, on the one hand, and deception, purposeful inaccuracy, and noncompliance, on the other (see detailed discussion in the next sections). These combinations are significant because they give equal weight to each node of the triad making elements appear interchangeable. Truth is equivalent to compliance, for example, but compliance is also equivalent to truth, and so on. In line with my overarching argument, Marston's experiments are not intended to discover truth or deception; they are designed to produce a deceptive consciousness that will produce a significant lying curve. Within the context of this first experimental protocol, the deceptive consciousness results from noncompliance and inaccuracy.

One could imagine—and we do indeed have the anecdotal narratives to ground such imaginings—that noncompliance could include actions that were outside the parameters of the experimental protocol. In later years, as Leonarde Keeler further developed Marston's technique, he often used a "card trick" to establish a baseline for his subjects and also prove the amazing reliability of his machine: the Keeler Polygraph. He would ask subjects to draw a card, examine it, and place it back in the deck of ten cards. Then, one by one, he would draw out the cards, asking subjects, "Is this your card?" Subjects were, as in Marston's experiments, instructed to "lie" by answering "no" to all of the cards, including the one they initially drew. Keeler would then examine the graphic record and, judging from the spikes in autonomic responses to the "lie condition," would be able to determine which card belonged to the subject. In Ken Alder's history of the lie detector, he notes that the woman who would later become Keeler's wife thwarted his card trick test, not because she was a great liar, but because she did not follow the initial protocol of Keeler's experiment: she never looked at the card she drew (Alder 2007, 83). Therefore, even she didn't know when that particular card was presented to her, her negative responses were never deceptions—to her, at least—and thus, Keeler could not find the lie and thereby identify her card. In this case deception took place, but it happened outside of the experimental protocols and could not be detected by the test as it was configured. However, Kay Applegate Keeler's "trick" does substantiate Marston's initial hypothesis about the deceptive consciousness: she did not express an attitude of deception in response to the only question asked ("Is this your card?"), and the record of her autonomic nervous system corroborates that fact.

To avoid similar loopholes, Marston's research became more and more elaborate as time went on. He moved away from word-association tests to overtly performative deception-based narrative and the literal performance of staged crimes. In the second role, subjects were asked to play the part of (false) witness. Marston tells us that "the subject came to the experiment as to an examination by a prosecuting attorney, resolved to save a friend who was accused of a crime" (1917, 124). In the context of the experiment, subjects had to testify about their friend's involvement in a particular incident. To do so, subjects were required to choose from two cards labeled "Truth" and "Lie" on which various facts had been written that indicted the subject's friend. If the subject chose the "Truth" card, she simply had to narrate the facts as they were presented. "This story was the *truth*," Marston explains; "it was the only account [the subject] knew of the affair, and he told it as such" (124). If, on the other hand, the subject chose the "Lie" card, she was asked to construct an alibi that jibes with facts corroborated by other witnesses, but which also exonerates her friend. False alibis are built, very explicitly, on already established truths: a subject who chose to play the false witness "proceeded, with these facts and the true story before him, to think out a consistent lying alibi" (124).

Again, we see the effect of a problematic triadic conflation; in this case, truth is equated with limited narrative choice and primacy of knowledge ("this story was the *truth*, the only account the subject knew") and with reiteration, in this case the repetition of an already assembled story. If, instead, the subject crafted an alibi—if she bore false witness— she was deemed to have "deceived" the experimenters, deception being equated with multiplicity of narrative choice, inhibition of knowledge, and iteration of a novel account. Truth telling is constructed as the base-line, as the primary story, the one from which a lie must diverge and be fabricated; indeed, storytelling itself is condemned through its associa-tion with deception.

Playing the role of mock criminal (a technique that is still used in contemporary studies of deception)[28] is an extratextual task that involves the perpetration of a staged crime, usually stealing, followed by an inter-rogation in front of a jury and an examiner. Unlike the false witness, the mock criminal is trying to protect and exonerate herself, not save a friend. William Marston's first published experiment illustrates the typi-cal protocol for the mock criminal. Each subject left the room with a sealed envelope. If, upon leaving the examination room, she opened the envelope, she was required to commit the crime listed therein and sub-sequently lie about it. If he chose not to open the envelope, he "did what-ever he liked for 10 min.," came back and gave a truthful account of his actions (1917, 125). Marston's experimental protocol at Camp Green-leaf was even more elaborately staged; he described it thus.

> *Method.* A. *Crime.* About 50 articles, each of some intrinsic value to a soldier, together with ten five-cent pieces, were disposed about a room on the second floor of the Psychology Building. The men were then instructed to enter said room, examine contents, and if they so chose, to steal and conceal upon their person one or more of said ar-ticles. If they chose to steal they must hide the stolen articles within the Psychology Building within 5 minutes after taking same; and in 10 minutes thereafter, they must take the stolen article out of the build-ing, convey it to their barracks, and there conceal it among their ef-fects. When examined, they were instructed to do their utmost to convince their examiners of their innocence. (1921a, 567)

I include this description in full to illustrate Marston's definition of "crime," which is, like the deceptive consciousness, both psychological and physiological. Similar to Luther Trant's theories about the marks of crime on men, Marston hypothesizes that the commission of a crime will leave a trace of fear and anger in the psychology of a suspect. For this rea-son, Marston did not maintain a record of physiological changes *during* the mock crimes. Instead, he recorded autonomic nervous system fluctuations as subjects later testified about their actions. However, he deemed the commission of a mock crime an important enough component to warrant an elaborate staging, one that he hoped would create a more prominent struggle between body and mind as participants enacted and reacted to the mock crime scenario.

Even under ideal laboratory conditions Marston was not always satisfied with the results he was able to attain. He made the decision, for example, to throw out an entire week of data from the 1917 experiment because "it was evident from the records that the subjects had disobeyed instructions, mixing truth and falsehood in such a way as to produce no clear consciousness of either, under the laboratory conditions obtaining" (1923, 392). When subjects failed to follow protocol, when they augmented "true" stories, or made a mistake in the details, they did not display the deceptive consciousness Marston sought to isolate. We know this because of the extensive individual subject analysis Marston provided in his experimental write-up. Using a combination of systolic blood pressure records and introspection, Marston explained how we are to interpret the data sets and graphs presented to us. Subject E, for example, "added several details to the 'T' story, but claimed to regard this just as 'telling it in his own words,' and introspected 'no excitement'" (1917, 141). Subject G, who adjusted the given 'T' alibi because he thought it "improbable," "felt throughout that he was correcting a mistake, and telling the *real truth* so that this record seems fairly listed as a real truth record" (147; emphasis added). This second example illustrates the distinction Marston wanted to make between deception as objective category and deception as an attitude or consciousness. Other subjects know they have made what Marston calls a factual "mistake," but they do not admit it and/or they attempt to deceive the examiner about their error. After one such "mistake" results not in a rise, but a drop in the systolic blood pressure record, Marston tries to explain it by saying that the "subject was conscious of having made uncorrected mistakes. Evidently whatever emotion accompanies *this idea* does not increase the b.p." (145–46; emphasis added). "This idea," that mistakes need not be corrected for—and are not, in fact, deceptions—complicated Marston's experimental search for the deceptive consciousness, while also revealing his own a priori assumptions. He did not term this kind of response "deception," nor did he include it in his definition of the deceptive consciousness. These minor classificatory differences cue us in to the possibility that Marston's protocols were recursive; they sought something that was already posited and defined: the deceptive consciousness.

## Pleasant Prevarication: From the Laboratory to the Field

Despite the need for tightly controlled protocols, Marston continued to believe that the benefits of a test and technology capable of detecting the deceptive consciousness reached far beyond the laboratory—they also embodied the possibility for profound social transformation. Particularly in his later texts, Marston argued that the technology could replace the violence and subjectivity of third-degree police interrogations, enhance interpersonal relationships, and create a more efficient society that was not plagued and burdened by criminals capable of deceiving investigators. However, to isolate the deceptive consciousness, to identify and measure it, Marston relied on laboratory protocols that, particularly in the case of deception, do not translate well from laboratory to field.[29]

First, the difference between lab and field appears particularly pronounced in lie detection as scientists ostensibly move from experimental subjects (the *antagonist, false witness,* and *mock criminal*) to criminal suspects. The first three types of liars I discussed are all products of the laboratory stage; their deception (and deceptive consciousness) is created in an artificial environment in which life and happiness are not threatened. Marston argues, for example, that deception is more dangerous and more deeply felt in real-life situations "in the actual court work, life and happiness might hang on the success of a deception, it is much more doubtful whether the whole situation during deception would be more pleasant than while telling the truth" (1917, 158). Yet, Marston's protocols assume that subjects' consciousnesses can be staged through particular behaviors and performances to represent—at least temporarily— the deceptive consciousness. Subjects literally perform a criminal act like actors on a stage, introspecting various emotional states. Conversely, the *suspect* is a product of the police/judicial system. She is the only individual who is considered a criminal suspect by a court of law,[30] and has already been charged with and arrested for a crime.[31]

We cannot overlook the fact that in Marston's experiments, there is a marked difference in motivation from consequence to reward. Because subjects are asked to lie, prevarication and the ability to fool one's interrogators take on positive connotations. One subject, Marston notes without a hint of irony, "made an excellent witness, telling very plausible complete lies" (1917, 136). Several subjects introspected[32] that they "found lying 'restful,' 'lax,' and 'pleasant'" (131–34). Others related that "deception was pleasant, exciting, 'tense emotionally,' with a 'feeling of excitement near the heart'" (151). Truth telling was, for many subjects, "uninteresting" (141). In short, subjects were invested in lying, but for different reasons than one might expect: the efficacy of their performance—not their lives and liberty—depended on their success. Indeed, if subjects succeeded in fooling the examiners and the jury, they were often allowed to retain possession of the stolen goods. "If they stole and yet succeeded in deceiving their examiners, they could keep the article stolen; if detected they must return same" (1921a, 567).

Marston attributes remarks about the "pleasant" nature of prevarication to his subjects' interest in the "whole proceeding, as an adventure is more pleasant than routine" (1917, 153); more specifically, he links this response to other emotional experiences, competitive gaming experiences in particular, noting that "men prefer bluffing at poker to playing a conservative hand, and the explanation would seem to be that excitement is more pleasant than quietude, any emotional experience being preferred, perhaps, to a purely intellectual activity" (158). In later articles, Marston directly compares lying in a laboratory to a "game" (1920, 79). This analogy implies that like other games, lying involves rules, a plan, and various techniques that could be learned and perfected. Indeed, the subjects who admit to feeling less invested in the experiment are the ones who choose truth most often. Subject H, for example, "took less actual interest in any sort of work than any other subject and for this reason both the number and quality of his 'T' records are significant. His lack of active interest, he introspected, led him to choose the truth 7 times out of 10, and the feeling persisted throughout these 'T' records" (1917, 147).

In addition to the problem of motivation, Marston introduces a relational requirement into his experimental protocol reminiscent of "interesting mixtures of truth and lying" I mentioned earlier: examiners need to know the truth before they can diagnose deception. This comparative relationship harks back to Marston's identification of the systolic blood pressure symptoms of deception that produce a significant lying curve. Marston's definition (and his subsequent and related assumptions that excitement is preferred to quietude, and deception is preferred to truth telling) sets in motion an elision of norm, baseline, resting state, and even default, with the "truth." This paradigm is repeated in contemporary neuroscientific lie detection experiments.

The basic principle at work here is that through experimental protocols, such as the mock crime scenario, scientists can construct truth as
known and singular or as a baseline against which deception can be measured. The truth can be characterized as known and singular when the
truth-statement is prepared ahead of time and is accessible to the researcher. In Marston's experiments, for example, "if the subject chose to
tell the truth, he turned over the "T" paper . . . This story was the *truth,* it
was the only account he knew of the affair, and he told it as such"
(Marston 1917, 124). Here "the *truth*" is defined not abstractly or philo-
sophically, but concretely, as "the only account" known to the subject. By
extension, lying is defined as the known suppression of "the *truth.*"

In the latter case, truth is defined as a baseline whenever a norm is established. Norms are generally established in one of several ways: subjects can relate a story known to be true, they can sit quietly before and
after an interrogation, or they can tell a lie that is known to require deception.[33] Marston establishes norms through the former two methods,
and we have already partially seen the latter in the ways that deception is
purposely invited into the laboratory. In the course of Marston's and subsequent experiments, the norm is generally established when subjects
are asked to sit quietly before and after questioning. This quiet period is
said to establish their baseline blood pressure, respiration, and heart
rate. Like the subject telling the truth who feels little emotion, the
record of a subject at rest, neither actively lying nor telling the truth, is
expected to register less intensely on the graphic record.

Yet, in the conclusion to Marston's first experimental write-up, he
makes the following recommendations that conflate truth and norm:
first, "two records must be taken . . . the story told during one record being truth within the knowledge of the examiner" (1917, 163). A similar
condition is instituted in the 1921 experiments when Marston insists
that we must choose "those cases where the blood pressure judgment as
to truth or falsity could be immediately checked" (1921a, 555). This in-
sistence not only returns the deceptive consciousness to the artificiality
of the laboratory, but it also compounds the idea that deception and
truth are relational. Without a record of a subject's "truth curve," the
significant lying curve and therefore its extrapolated concept (the deceptive consciousness) would be implicitly more difficult, if not impossible, to identify. In Marston's second recommendation, he urges that
"above all [the b.p. record during testimony] should be compared
minutely with the record known to be symptomatic of that individual's

consciousness while telling the truth" (1917, 163). Disregarding, for a moment, the individuation of this last statement that contradicts Marston's search for a uniform marker of deception, we are left with another startling revelation: the mechanical record associated with a truthful consciousness serves as a baseline or norm for any deviation, here defined as "an indicator of deception" (163). I analyze this problematic elision between truth and norm in chapter 5. Suffice it to say here that Marston's experimental protocols were the first and most influential in establishing this relationship. His experiments created and defined not only a working object (the significant lying curve) but also a new language through which to conceptualize the evocation of the deceptive consciousness in laboratory experiments. He gives himself, and later generations of experimenters, the right to evaluate performances, even to the paradoxical point of telling us that "objectively," one subject's alibi "was a wild lie" (134).

After William Marston's techniques were deemed inadmissible by the *Frye v. U.S.* case of 1923, the young psychologist turned his attention away from the laboratory and toward the popular press. Unfortunately, he found little acceptance in this realm either. When Marston's book-length treatment *The Lie Detector Test* appeared in 1938, his narrative of unparalleled results, sensational reliability, and criminal conquest drew frustrated reactions from several fellow scientists. Fred Inbau was the first to respond. He declared in his book review that "in making such sweeping statements Marston forgets that in his book he is supposed to play the role of a *scientist* and not that of a popular magazine writer, or newspaper reporter, or special guest on some advertiser's radio program" (1938, 305). In some respects, Inbau's assessment was prescient: William Marston is primarily remembered for playing the role of a showman in a series of Gillette advertisements in 1937 and 1938, a two-page spread in *Look Magazine* in 1938, a popular experiment testing the amorous inclinations of blondes, brunettes, and redheads chronicled in the *New York Times* ("Blondes" 1928), and creating the lasso-toting Wonder Woman in 1941. What has been ignored in the historical and scientific record is a scholarly analysis of Marston's role as a traditional experimental scientist.

Such cultural and disciplinary divisions did not hinder Marston's career: he not only challenged the divisions between scholarly fields but, as a persona, moved fluidly between the laboratory and the field. In fact, *The*

*Lie Detector Test* itself represents a mixture of public salesmanship and insider explanation: the final chapter, "Practical Suggestions on Lie Detector Technique" (which makes up a third of the book), is intended "For Users of the Lie Detector Test only" (1938a, 147). Despite (or perhaps because of) his "unscientific" dabbling, Marston made a name for himself with both audiences. As Molly Rhodes notes, within both the academy and the public spheres, "his writing circulated as wholly legitimate academic science, having the weight behind it of a Harvard-trained PhD licensed to practice and teach law and psychology" (Rhodes 2000, 99).

Marston's research brought applied psychology, emotional inscription technologies, and lie detection to the public consciousness. He played an instrumental role in developing techniques (discontinuous systolic blood pressure measurement, mock crimes), working objects (the significant lying curve), and concepts (including the deceptive consciousness) that would affect lie detection protocols for years to come. Along with comic books, he wrote textbooks, clinical manuals, and laboratory reports. As we saw in chapter 1, Marston was also the first scientist to bring lie detection evidence to the courts. Marston may not have invented *the* lie detector, but his life's work was mutually imbricated in the genesis, genealogy, and survival of lie detection as technique and technology, scientific instrument, and "psychological medicine" for social betterment (1938a, 15).

In this chapter, I analyzed Marston's concept of the deceptive consciousness with particular attention to the foundational assumptions informing its definition and use in laboratory experiments. We saw that Marston's measurement techniques emerge from and dovetail with a long history of emotional inscription technologies, yet his evocation of lying in the laboratory required the construction of a new interpretive architecture. Tracing the generation and history of the deceptive consciousness also provides a bridge between the early literary and legal marketing of lie detection via *The Achievements of Luther Trant* and Hugo Münsterberg's applied experimental psychology work and contemporary brain-based experiments on deception using fMRI and EEG. In all three cases, scientists are looking to find "the marks of crime on men." As a theory, Marston's concept of the deceptive consciousness provides a concrete means through which to examine the experimental assumptions that undergird this relatively modern pursuit.

# Thought in Translation

## Reading the Mind in Science and Science Fiction, 1930–50

> In a very true sense psychology contains all the possibilities of fiction and can become as fascinating.
>
> —DAVID SEABURY, *Unmasking Our Minds*, 1924 (xxii)

In October 2003, PBS and *Wired Magazine* featured brain-based lie detection on their new show *Wired Science*. The show and the imaging technology it featured were advertised with the caption "We've got mind reading down to a science." The unspecified *we* is reminiscent of the ever-vigilant Esper Police in Alfred Bester's *The Demolished Man* (1951/1953) or perhaps the Thought Police of George Orwell's *1984* (1949). But in the *Wired Magazine* advertisement, reading the mind through brain imaging suggests that science has surpassed science fiction in that there is an objective, interpretable correlation between anatomy, physiology, and thought. Through the technologies of brain imaging, the advertisement claims, we can finally visualize not only the brain but the mind itself. As with Hugo Münsterberg's mental microscope, the foundational assumption is that the mind (though not inherently material) can be made so through technical intervention. And forget about hiding from or evading the searchlights of science. Below the caption, the ad copy reads, "Beating a polygraph test is one thing. Beating a machine that can actually read your thoughts is another."

As this advertisement illustrates, popular representations of brain

imaging often envision the technologies as not only objective and invasive but also capable of translating electrical activity or blood oxygenation levels into meaningful messages. In short, we imagine these technologies could do what the ad copy promises: read our minds. In recent
years, scholars have addressed representations of brain imaging in the
media with specific attention to the way visible models are often seen as
more objective and true than other types of data (Galison 1997; Daston
and Galison 1992; Joyce 2008); the tendency for and consequences of
using imaging and visualization as a means to understand human behavior (Dumit 2003); and the elision of brain and mind, or the "mind-in-
the-brain" (Beaulieu 2004). In this chapter I illustrate a complex genealogy for our popular perceptions of brain imaging technologies, one
strand of which goes back to the development of a literacy for mechanical mind reading in the sciences and science fiction of the 1930s to the
1950s. During this era telepathic experiments were recognized as academic science, human electroencephalography (EEG) produced
graphic images of the brain's electrical activity, and fictional narratives
imagined machines that could read the mind, projecting thoughts as
sound and image.

Examining these earlier conceptions of mechanical "mind reading"
in multiple disciplinary sites reveals a persistent belief in a psychophysiological literacy that constructs the mind as, at best, a transparent
medium and, at worst, a text to be translated and interpreted by a set of
expert technicians. When this literacy was first constructed for and by a
lay audience, as it was in the sciences and science fiction of the 1930s to
the 1950s, its simplified, mechanized vision helped to shape perceptions
of and expectations for what would become the sciences of brain imaging. As we will see in chapters 4 and 5, the legacy of mind reading continues to inform the scientific and journalistic representations of contemporary lie detection. In this chapter, I address the construction of
several ideological assumptions about the mechanics of mind reading:
first, that thought and energy are essentially interchangeable or transmutable, if not one and the same; second, that thought can be translated
rather transparently from energies into recognizable sounds and images
and, moreover, these thought pictures are representative of one's character; third, that there is an anatomical and psychological space (the hidden lower brain) that houses dangerous, primitive thoughts; and, finally,
that thought can serve as material evidence in a criminal prosecution.

To make my case, I take seriously Susan Squier's assertion that "science and literature are more like each other than they are different, not only because both operate in culture and society to produce subjects and objects but also because both fields have come into being through a crucial act of institutional self-creation: the creation of a disciplinary divide between scientific and literary knowledges and practices" (2004, 31). Science and literature are not distinct enterprises, even though they often appear to be discrete because of the careful boundary work that continually reifies their borders. One way to unhinge this binary is to find and analyze the moments when science and literature create a common object or subject: in this case, a technology for and a literacy of mechanical mind reading. To do so, I examine not only the publications of scientists, including Duke parapsychologist J. B. Rhine; chemical engineer René Warcollier; psychotherapist and psychiatrist Jan Ehrenwald; and physician and psychologist Hans Berger; but also several science fiction stories and popular nonfiction volumes published between the 1930s and the 1950s, including Upton Sinclair's *Mental Radio* (1930) and a host of self-help books. I turn specifically to stories like "The Thought Translator" (1930), "The Thought Stealer" (1930), "From the Wells of the Brain" (1933), "The Ideal" (1935), and *The Demolished Man* (1951/1953), because these narratives imagined what science could not actually produce: mechanical mind reading that translated thought into sound, image, and matter. These fictional accounts did not publicize the instruments of psychophysiology that we saw in chapters 1 and 2; instead, they constructed plausible lie detection machines based on a mixture of telepathy, EEG, and thermodynamics. In science fiction we can find the earliest records of the ideological assumptions that undergird popular references to and assumptions about brain imaging as a science of mechanical mind reading.

## Making Thought Visible

Between the late nineteenth century and the early twentieth century many things that were once invisible became visible[1] via the invention, modification, or employment of various machines. Small units of matter, including germs, hormones, and neurons, were seen (and named)[2] for the first time using powerful microscopes; X-rays, which depended on the Crooke's tube, showed us the interior of the body without the need for invasive surgery. Forces that could be detected, even if they remained unseen, were also made concrete thanks to theories about thermodynamics, which affected the ways science conceived of the relationship between matter and energy (Clarke 2001; Smith 1998). Although unseen, phenomena of light, heat, electricity, and magnetism were also reconceived of as interchangeable forces.[3] "The new language of energy," notes Crosbie Smith,

> was symptomatic of a series of profound conceptual shifts which resulted in a whole new scientific vision, with accompanying changes in scientific practice, quite unlike anything that had preceded it. Although fundamentally mechanical in nature, the universe would now be understood neither in terms of action-at-a-distance forces nor in terms of discrete particles moving through void space, but as a universe of continuous matter possessed of kinetic energy. (1998, 2)

Matter was energy, and energy, matter. Even if our known senses could not recognize these unseen energies, machines operated by specialists could harness and visualize their potential. Radios broadcast voices across space (Squier 2003), and telegraph wires translated language from the air (Otis 2001), while physiologists were busily recording the body's own concealed capacities for work and emotion (Rabinbach 1990; Dror 1999a and 1999b).

Conceptualizations of thought in the early twentieth century were affected by these major movements in science that can best be characterized as the fragmentation of the body (Armstrong 1998; Burnham 1988),[4] the mechanization of vision (Dror 1999b, 2001b; Ward 2002; Crary 2000), and thermodynamic theories of energy (Clarke 2001; Smith 1998). Thought, that once ephemeral phenomenon, was reconceived of as part of this matter, and likewise composed of particles. By leveraging understandings of energy, matter, and fragmentation, the brain and the mind were conflated, and thought became transmutable, transferable, and translatable, given the proper protocols and equipment. The emergent neurosciences and the burgeoning discipline of psychology each defined different units of matter and energy in the brain: the neuron and the psychon, respectively (Littlefield 2010).

Parapsychological research and human electroencephalography experimentation, which were products and producers of the new scientific theories of the particulate, the visual, and the energetic, would both make it their disciplinary business to translate what they presumed to be the energy of thought into material form. Parapsychologists, such as J. B. Rhine, recorded the transmissions of thought in academic laboratory experiments. Indeed, Rhine's work at Duke University made parapsychology (telepathy, extrasensory perception, and telekinesis) a legitimate field of academic study. Meanwhile, psychiatrists, such as Hans Berger, captured and recorded brain waves in graphic form. Berger's work on human electroencephalography at the University of Vienna, which began in the 1920s, spawned nearly a decade of debate and incredulity from the scientific community. Unlike J. B. Rhine, Berger found academic validation not in his own laboratory but in the laboratory of two American scientists who, in the mid-1930s, confirmed Berger's original findings. As both approaches to thought energy are imbricated in the ideological assumptions about mechanical mind reading I analyze in the science fiction of the 1930s to the 1950s, I would like to take a moment to explain telepathy and EEG here.

Scholars have thus far provided detailed analyses of telepathy's emergence (Luckhurst 2002), its reputation among the natural sciences (Millett 2001), and its relevance to scientific and literary movements including "vibratory modernism" (Henderson 2002). However, none of these analyses attempts to explain telepathy's later history, including its turn to the visual between the 1930s and the 1950s. Understanding this turn to the visual legitimates telepathy as a key component in the genealogy of popular conceptions about brain imaging, including their ability to read our thoughts by making them visible.

The study of and belief in human telepathy[5] had long been motivated by efforts to prove that thought was both a transferable and a visible phenomenon. In a typical telepathic experiment, one person would concentrate on an image, a word, or an experience in an attempt to transmit that object or feeling to another person, often termed a percipient. The receiver, or percipient, would record whatever images s/he saw, usually on paper. Distance was not a factor, nor were material barriers such as walls. Parapsychologists around the world relied on images as the best transmissions for telepathy. The work of academic scientists such as J. B. Rhine[6] and René Warcollier and amateur scientists such as Upton Sinclair depended upon the image as a means to validate the telepathic experiment.

In both Upton Sinclair's *Mental Radio* (1930) and René Warcollier's *Mind to Mind* (1948; English translation, 1963),[7] telepathy is demonstrated through a series of correspondences dependent on visual records sent from telepathic agents and received by telepathic percipients. Prefaced by Albert Einstein and followed by an addendum by Walter

Franklin Pierce (research officer of the Boston Society for Psychic Research), *Mental Radio* is a "record of experiments, conducted in strict scientific fashion" that includes reproductions of hundreds of telepathically received images (1930, 23). The images themselves (see figs. 1–3) are rather rudimentary and inexact, as Sinclair readily admits; but, he argues, the received images do arguably share elements with the original drawings.

Drawings, like those in figure 1, do not have to be exact to be considered successful transmissions. Indeed, Jan Ehrenwald, a psychiatrist and telepathy enthusiast, notes that "Dr. R. H. Thouless 'emphasized that symbolic representation as opposed to photographic likeness of reproduction is a fundamental feature of telepathy'" (1948, 62). In any case, telepathy was thought to function by transmitting thought energies (that had been concentrated on a visual image) through the air to a receiver who could record them.

In his own work, René Warcollier explains the transmitted and received images' discrepancies in material terms that echo the major concepts of fragmentation and thermodynamics mentioned at the beginning of this section.

> The telepathic image is not transmitted in the same way as a wireless photo. The image is scrambled, broken up into component elements which are often transmuted into a new pattern. It seldom arrives complete and organized. A telepathic image resembles somewhat a chemical molecule. The original molecule, the target, decomposes into elements. Some of these elements are received and are recombined into a new molecular structure. (1963, 29–30)

Although Warcollier is reticent to admit any kinship between the wireless photo and the telepathic transmission, he compares the telepathic process to the fragmentation and transmutation of matter—decomposed and recomposed—at the atomic level. The forces at work are dependent on the personal and cultural baggage each individual brings to the telepathic experiment: "The telepathic percipient is not an automaton. His consciousness, in response to the latent image, does not operate like a telephoto machine or a piece of photographic printing paper on to which the negative has been placed" (55). What the percipient sees "is dependent upon the forces within himself. In this way, he reveals the structure of his own personality" (56). Like Hugo Münsterberg and William Marston, Warcollier believes that emotions and other psychic

On the following, also, no comment was written:

Figs. 1, 2, and 3. In these three images from Upton Sinclair's *Mental Radio* (1930) you can see the results of several telepathic experiments. Each image consists of two parts: the original visual that was transmitted (an abstract symbol and a palm tree; in the middle a cat chasing a toy; and on the right, a foot in a roller skate) and the visual image that was received by the percipient. In each case, the received image shares something in common with the transmitted image, even if it is a transmuted version of that image, as it is in the case of the roller skate/animal on the right. Material excerpted from the book *Mental Radio* by Upton Sinclair, reprinted with a new publisher's note by Russell Targ, with permission of Hampton Roads Publishing c/o Red Wheel/Weiser, LLC, Newburyport, MA, and San Francisco, CA, www.redwheelweiser.com, 800-423-7087.

phenomena imprint themselves on our material, physical bodies as well as on the energies we transmit. As we will see in the fiction, a minor but recurring ideological assumption about mechanical mind reading is that it both records and infiltrates beyond the emotions that affect our thoughts and behaviors.

Moreover, as we shall see more clearly in an analysis of the hidden lower brain, psychic researchers working between the 1930s and the 1950s often emphasize that telepathy taps into the symbolism—and primitivism—of our instinctual minds. In his foreword to Warcollier's book, Emanuel K. Schwartz argues that unlike graphic inscriptions, which purport to automatically and faithfully replicate physiological phenomena, "telepathic impressions have a quality of unconscious symbolism, of the eternal symbols found on all levels of development of man, at all stages of maturation, and in all cultures" (Warcollier 1963, 5).[8] Warcollier also suggests, for example, that "we are dealing here, it would seem, with mental powers that suggest an earlier stage of human development" (39) and that "the percipient in telepathy is like a child" (47). Scientific and fictional representations of telepathy suggest that the phenomenon taps into a Jungian collective unconsciousness—an inherited constellation of archetypes that help determine our personalities. Carl Jung, the Swiss psychiatrist who developed analytic psychology, was, in fact, a proponent of parapsychological phenomena, including telepathy.[9]

The rhetorical efforts to explain telepathic transmissions as material phenomena and the use of drawings, continued to mark telepathic events as objective and scientific, despite the turn to symbolism.[10] Indeed, telepathic transmissions shared much in common with the graphic technologies used to record other energetic waves. At its etymological root, the suffix *graph* (used to designate inscription technologies such as the plethysmograph, sphysmograph, and cardiograph) identifies instruments that "could 'write' or 'draw'" (Brain 2002, 159). Both graphic technologies and telepathy depended on a similar experimental medium: drawings, be they penned by the stylus of a machine or one held by a human hand. In *Mind to Mind*, Warcollier notes that "in most of our experiments we used drawings, as did other investigators, because they allow more precise check and control than do thoughts and verbal ideas" (1963, 26). These "drawings," argues Emanuel K. Schwartz in his foreword, "speak for themselves" (6). His description mirrors language used to depict graphic inscription technologies. As Bruce Clarke and

Linda Henderson note, mechanical inscriptions "were not gestural strokes of self-expression, but rather depersonalized lines associated with automatic recording apparatus and scientific diagrams, the 'language of the phenomena themselves'" (2002, 154). In both telepathy and other graphic technologies, nature appears to write itself through the medium of human hand or stylus.

During the same era in which telepathy gained academic validation, Hans Berger—who often lectured on the merits and possibilities for telepathy—developed a means to measure and record the electrical impulses of the brain through the intact skull, a technique that would become known as human electroencephalography. Hans Berger's pioneering experiments, which he undertook between 1910 and 1938 and first reported in 1929,[11] share the assumption that energies can be materialized and recorded. He "was convinced that mental processes involved energy transfer in the brain and must therefore be subject to the law of conservation of energy," and that "through conscious experience and the conversion of energetic processes in the cerebral cortex associated with them, the central nervous system is continuously being restructured and psychophysiological processes are laid down *in material form*" (Gloor 1994, 254; emphasis added). Berger's most successful technique, the one that led to EEG, involved the application of electrodes to the scalp. The electrodes measured the electrical potentials of the human brain; the resulting rhythms were recorded by an oscillograph. While Berger's experiments were initially ignored and/or met with skepticism, their value was finally realized when EEG was replicated by two Americans, E. D. Adrian and B. H. C. Matthews, in 1934.[12]

Hans Berger's experimental work allowed for the visualization of the brain's electrical impulses; but his imaginative vision[13] was to create a direct translation of human brain function: a "*Hirnspiegel*" (Berger's Diary, November 16, 1924; quoted in Borck 2001, 583). Berger's choice of the term *Hirnspiegel* can be and has been translated two ways, as cerebroscope and as brain mirror. The choice of the latter translation by historians of science Peter Gloor and David Millett emphasizes not the literal German translation (*hirn:* cerebral and *spiegel:* scope/speculum) but Berger's excited prose and hopeful vision best captured in his personal prose.[14]

Although Berger was fairly reserved in his public scientific writings about EEG, his enthusiasm was picked up and extended by the popular press. "In contrast to Berger's own discretion, the press did not share his hesitation and did not present the EEG as a discrete gesture toward the psychological, but rather as its deciphering. More than simply a means for recording brain activity, electricity had turned, according to these articles, into the language of the operating brain" (Borck 2001, 584). Take, for example, an article that appeared on the front page of a special to the *New York Times* in 1935 that begins with the headline "Electricity in the Brain Records a Picture of Action of Thought. Currents generated by mental processes are transferred into visible patterns written by mind-reading needle which records variations in cerebral adjustments" (Lawrence 1935, 1). The primary literacy of this mechanical mind reading is visual: "visible patterns" and "picture." It is also a more holistic and idealistic description, akin to Berger's own desire for a "brain mirror."

Literary representations and scientific utilizations of EEG (and EEG-like) technologies tend to reinforce and/or rely on Berger's original conception of a "brain mirror."[15] As I argue in the rest of this chapter, between the 1930s and the early 1950s, science fiction would take these theoretical trends in imaginative new directions, combining telepathy, EEG, mechanical vision, fragmentation, and thermodynamics. In fiction, thought was rendered visible through a wonderful array of thought-projecting helmets, rays, projectors, and translators that took full advantage of—and imaginative license with—scientific theories.

## The Self-in-Translation

At the same time the sciences of telepathy and EEG were constructing techniques to record thoughts, the fiction of that era begins with the supposition that thought is energy and is therefore material and transferable; however, instead of translating this energy into graphic outputs (as in EEG) or symbolic images (as in telepathy), thought translation in science fiction is typically accomplished through machines that transmute thought energy directly into sound and spoken language, light and pictures. In some cases, as in Paul Ernst's "From the Wells of the Brain" (1933) or Stanley Weinbaum's "The Ideal" (1935), thought energy is transmuted directly into matter. In the latter, for example, we are told that Professor van Manderpootz's machine can change the energy particles of thought (which he terms "psychons") into matter just as "a Crookes tube or X-ray tube transforms matter to electrons" (Weinbaum 1974, 225). In other cases, as in Merab Ebertle's "The Thought Translator" and Alfred Bester's *The Demolished Man*, thought energy is translated into sound, image, and in the case of the latter, textual word patterns.

Such mechanically mediated visualizations make thought seem not only accessible but translatable: this energy can be readily and transparently moved from one medium to another. I use the term *translation* here and in the title of the chapter because I want to stress the role of interpretation that is both explicit and implicit in the act of *reading* the mind. As I illustrated in chapter 2, new applications of technology demand the creation of new interpretive language. Translating thought energy into preestablished representational systems, such as sound and image, cannot assume a one-to-one correspondence; such a shift involves the transmission of meaning-full messages that have to be captured in a new medium. Within the science fiction of the era in question, the initial translation of thought energy, as well as any interpretation of the newly translated material, appear transparent to the machine and the scientist involved.

Given its ostensibly material form, thought becomes one more facet of the physical body, and, like the body, thought becomes implicated in discussions of identity and character (Cole 2001; Thomas 1999; Gould 1996). Instead of the adage "you are what you eat," it could be said that "you are what you think." As individuals are made privy to the visualizations of their thoughts, internal and external conceptions of their character both impact and are impacted by the translation process.

In recent years, the issue of the brain and the self has been addressed by Joseph Dumit, who argues for the "objective self" as a means to understand interactions between science, the media, and our conceptions of self. He draws examples primarily from the brain imaging technology of positron-emission tomography (PET) scans and its representation in and by the media. According to Dumit, the "objective self is an active category of the person that is developed through references to expert knowledge and invoked through facts. The objective self is also an embodied theory of human nature, both scientific and popular" (2003, 164). He goes on to explain about the process.

> Objective self fashioning calls attention to the equivocal site of this production. . . . From one perspective, science produces facts that define who our selves objectively are, and which we then accept. From another perspective, our selves are fashioned by us out of the facts available to us through the media, and these categories of people are, in turn, the cultural basis from which new theories of human nature are constructed. (164)

In short, the objective self and the process of fashioning an objective self are dynamic and agential within a particular matrix of power. Selves are both products and producers of human natures and human types that are propagated through science, the popular press, and individuals.

The materiality of thought, as it appeared in and impacted the science and science fiction of the 1930s through the 1950s, can provide a supplement to Dumit's theorization of the *objective self*. Here, I suggest a new term and concept: the *self-in-translation*. The self-in-translation builds from Dumit's "objective self" to highlight the translation process inherent in mechanical mind reading (as it was introduced between the 1930s and the 1950s) and to suggest that this translation process makes thought into another facet of the physical body, one that can be monitored, assessed, and implicated in assumptions about an individual's character. In this respect, the mind became a site of knowledge about the individual not because of its Cartesian separation from the body, nor because of its elision with the brain, but because thought was conceived of as a particulate, material substrate.

Between the 1930s and 1950s, in particular, science fiction helped produce a self-in-translation that valued the characteristics of restraint and self-control, aspects of which are made abundantly clear in Merab Ebertle's "The Thought Translator," a short story published in *Science Fiction Series* (1930).[16] In this narrative, scientist Alfred McDowell renders thought visible through what he calls "my new discovery—that of correlating the oscillation of thought, which is also ether-carried, and converting it into pictures and sound" (16). Using this device, he helps to mediate several misunderstandings between his friends: a judge, Justice Crowell; the judge's unrequited childhood sweetheart, Alicia Wentworth; and, notably (given the focus of this chapter), a journalist, Dick Green. After the three meet in McDowell's laboratory and have their thoughts voluntarily and involuntarily translated, they are equipped to resume lives that are founded on transparency: the judge now understands why Alicia refused his proposal decades earlier; Alicia understands how she was duped by her best friend into refusing the judge's hand; and the journalist learns what kinds of information are appropriate for distribution to the newspaper-reading public.

The process of thought translation in this story draws attention to the dynamism of thought and its potential illegibility within established systems of representation. We are privy to three types of translation: garbled sounds, language, and moving pictures. When McDowell's machine translates thoughts into sound, they are represented both as noise and as clear speech. The former characterization reflects the incongruity between thought and language. The narrator's thoughts are initially "a jargon of sound with scarcely distinguishable words, soft and compelling" (6). Later, the judge's thoughts are described in similar terms: "Tumbled words threw themselves softly through the air. Meaningless phrases piled one upon the other" (8). When thoughts are translated into pictures and projected on the south wall of the laboratory, Dick mistakes them for bad cinema, commenting that "the moving pictures were so poor. They came and went, faded, and then shot forth into clear images. The faces shifted expressions quickly. First there were children and then grown men and women" (11).

Importantly, one of the main differences between auditory and visual translation is the relative struggle for control between the character and his/her thoughts. When thoughts are articulated in oral language, the narrative tends to reiterate what telepathy, EEG, and thermodynamics took for granted: that there is an essential connection between thought and body, energy and matter. This interlinking of mind and body echoes what Hugo Münsterberg theorized at the turn of the century: that no matter how much self-control is exerted, the body will betray one's thoughts. As thought is transparently translated into language, two voices can be heard: that of the character's conscious voice and that of his/her heretofore unspoken thoughts. The struggle to silence the second voice creates mental and physical conflict for the characters who do not want their secrets exposed. When thoughts are articulated with clarity, the characters each turn to their bodies for answers and abatement. When the judge's thought is translated, the voice is so distinct that "he started back, amazed, puzzled—feeling his lips as if to prove to himself they were not moving" (8). Upon hearing his own thoughts, the narrator attempts to distance his self from them, explaining that "it was not my voice, it was my thought that filled the room" (6). Indeed, this vocalization of thought is eventually termed "appalling" by the narrator who begins to "plead with [McDowell] both through word of mouth and vocalized thoughts. 'McDowell, for God's sake unloose me from this . . . this . . . I can't stand it" (7). As a final measure against self-exposure—one that relies on assumptions about the body's role in self-incrimination—the narrator explains, "I shut my lips tightly" (7). Unfortunately, thoughts are not so easily silenced.

Because the incriminating evidence (i.e., thought) appears to voice itself in a kind of automatic translation—to be out of the subject's control—the mechanical thought translator exposes fears about the potential shame of unwarranted disclosure. Early on, the narrator expresses an ambiguous relationship with his thoughts, claiming, "My thoughts were attempting to take form" (Ebertle 1930, 6), they "were making themselves audible. I listened to my own mind. With my own ears I heard my thoughts expressed simultaneously with their conception in my brain" (7). The narrator's thoughts are separated from him in certain key respects—it "was not his voice"—even though when he subsequently claims them as "my thoughts," they appear to be independently taking form. In any case, the narrator finally admits, "I was ashamed of the way in which my thoughts took form. There is much of the child in the man's mind which he endeavors to keep from the world, and does—fairly successfully. But the child made itself known this time. My thoughts whined at times" (7). The shame is not in exposure, but lack of composure and control. He is ashamed because he cannot "keep from the world" the child inside the man. His thoughts reveal him to be something that he wishes to distance himself from.

Unlike the mechanical vocalization of thought, which often reveals the spontaneous and ill-formed musings of the mind, the mechanical visualization of thought appears much more within the control of the individual. In part this is because memories have been crafted and recrafted over time. Self-control implies a different kind of discipline: the choice to be an active participant in the shaping of your thought pictures for others to consume. When Crowell voluntarily undergoes the visualization of his memories, McDowell asks if he would "explain the pictures as they are thrown on the screen? This will make it easier for you, I think, than the use of the talking thought-translator. Then too, I believe that your effort at explanation will tend to co-ordinate your thoughts so that they will be pictured more definitely—with greater clarity" (11). The judge's thoughts are clearly a reconstruction (rather than an exact replica) of an earlier time, as he even notes, "It seems difficult to recall myself as I was. Have you ever dreamed, McDowell, that you were back in the first grade at school and were endeavoring to take a seat—yet found yourself too large? This object that looks something like a cross between a boy and a man is myself, Richard Crowell" (12). Here, Crowell not only acknowledges his thoughts (and their shortcomings), but he also identifies with them. He is not ashamed of them; rather, he is amused and engaged in their construction and illustration.

Indeed, the visual thought translator (much like the Idealizer in Stanley Weinbaum's story "The Ideal") affects the very contours of thought—

and therefore images of one's character—that it purports to represent. Translation, in other words, is not a process of complete transmittal from one system of representation to another. When visualized, thoughts take shapes that affect and are affected by the self-in-translation; they can be augmented, manipulated, and changed. In "The Ideal" the narrator notes, "The very fact that I had seen an ideal once before had altered my ideal, raised it to a higher level. With that face among my memories, my concept of perfection was different than it had been" (1974, 231). Judge Crowell experiences a similar effect; as he sees and narrates his projected thoughts, he notes, "The picture on the wall adds to the depth of my mental vision" (12). More significant, he tells his audience, "You have no idea how this picturing upon the screen seems by some uncanny method to clarify my thought and make me see more vividly than is possible mentally" (14).

Concomitant with the era obsessed with the visualization of thought and an era desirous for social reform through lie detection (see chapters 1 and 2), subsection titles within "The Thought Translator" draw attention to the connections between thought, confession, truth, and the accessibility of thought. In the context of many of these science fiction stories, the process of translation is meant to be revelatory, reformative, and redemptive, demanding confession and evaluation. And, as I pointed out in chapter 2, confession can be as nerve-wracking as deception. The journalist must confess the wrongdoings of one friend to another; the judge must confess his true love lies with his childhood sweetheart, not with the woman he has chosen to marry; and Alicia must realize that she has been denied true love through the treachery of a close friend. Sections of the story are aptly titled "An Involuntary Confession," "Crowell is Caught," "A Confession," and "The Truth Is Discovered."

Not inconsequentially, human judges, such as Justice Crowell, are shown to be wrong in their evaluations, while scientists (via their machines) are figured as experimentalist confessors, the ultimate arbiters of misdeeds. McDowell is literally and figuratively at the controls of the machine and all it represents. He chooses who to direct the machine toward, establishes what information should be public, and passes judgment on those he examines. When he decides that the judge is the next best candidate for thought translation, McDowell authoritatively tells him, "'I think, my dear friend, that you need to pay just a trifle for your statements in regard to this young man—pay a little for it.' McDowell's hand was on the dials again" (8). When Dick's thoughts are translated for the first time, he feels explicitly objectified by McDowell, who regards him "as though he were gazing upon a specimen through a microscope" (7). And after a few moments of intensive attention from the machine, Dick explains, "I had the sense of being before the Seat of Judgment and having the Power of Powers unveil my misdeeds to the hosts of the world" (6). The journalist is not only evaluated but exposed, "unveiled," for all of his "misdeeds." Despite this description in which the machine divulges Dick's secrets, McDowell has already patronizingly assured him, " 'Surely there is nothing in that brain of yours that you would really fear to have your old friend, your best friend, know" (7).

In hopes of combating any self-revelation, each character who comes in contact with the thought translator attempts to thwart its invasion by the same means that subjects have attempted to thwart the polygraph: through self-control. Their efforts jibe, more specifically, with one of the foundational assumptions of lie detection, that it takes extra effort to deceive (oneself or another person). Or, put another way by McDowell, "you will find this much less strenuous if you let yourself go and do not attempt to keep your thoughts from me" (7). Each of the characters in the story expresses the strenuous nature of their endeavor to deceive both the machine and themselves. Upon hearing his thoughts reveal himself, Dick exclaims, "I must take my thoughts in hand—must guard my ideas. And I heard myself thinking: 'I must be guarded, careful' " (6). Taking one's thoughts "in hand" is a somewhat unexpected expression. One's physical body becomes one's only means of controlling the expression of thought; this control is not concerned with the erasure of thought but, instead, its expression in language. "Must concentrate so that my thoughts shall not be uttered," Dick thinks out loud. "This flashed through my mind" (6). Once Judge Crowell's love interest becomes vulnerable to the machine, she too muses, "I must be very careful what I think" (17). Dick interprets this as being "strong-minded": "A strong-minded woman, this, I thought, for the picture which came upon the screen showed no whit of agitation, so quickly had she gained control of herself. Strong-minded, indeed, to be able to throw her thought into the channels that she desired" (17).

Strong-mindedness implies a certain mastery over the self-in-translation, including the inhibition of less desirable responses. Control, then, has to do with emotional agitation and outward expression. Such self-control shares much, perhaps, with the history of and theories about human inhibition (Smith 1992). Inhibition, notes Roger Smith, is a term that draws attention to what has been the nature/culture paradox: do we naturally inhibit certain behaviors because of a biological imperative and/or are we trained to repress certain behaviors, attitudes, tendencies, and even thoughts due to cultural constraints? Is inhibition necessary—even unavoidable—or is it an expendable act that is required in various social contexts? "The Thought Translator" muddies these waters further by posing questions in terms of self-control, while also illustrating both the amorphousness and repetitive structure of thought itself. Thought appears to be both amorphous and supremely structured, chaotic and inhibited, its basic form upheld by "self-control."

Ultimately, Dr. McDowell opts not to unleash his machine on the world at large, not because the machine has any particular failing, but because it would be all too powerful and accurate in its translations. Dr. McDowell explains all of this to the narrator as a rebuke against the journalist's impulse to share this machine with the world. The thought translator "is too tremendous a thing to be launched into the civilization of the day," notes the scientist.

> People could not endure the discoveries of human nature which I would precipitate into their midst. Young man, I am not convinced that it will be of such inestimable value for us to unlock the secrets of our thoughts. It is in the silences of our minds that we wrestle with angels through the night. Our characters are molded through our victories. Would we be benefited in the end, would the race as a whole, be aided by the revelations which would come to us? (18)

McDowell assumes that human nature could be discovered through the machine's translations; that we could see the molding process that usually takes place in the dark recesses of our minds. If, as McDowell does acknowledge, "our characters are molded through our victories," they are also molded through our interactions with science, technology, our previous conceptions of self, the myriad dynamism of human nature, and the ways each of these is constructed in particular cultural contexts.

### The Hidden Lower Brain

Dr. McDowell's characterization of the mind's inner recesses notably describes an individual battle with angels, rather than demons. In his ability to imagine the nobler character of thought, the doctor's vision jibes well with several other narratives of the era, including Stanley Weinbaum's "The Ideal." However, McDowell's machine examines only the conscious levels of thought, the thoughts we can articulate but often choose to keep quiet. According to American science fiction of the early twentieth century, the literate practice of mind reading demands that we also examine those recesses of mind—and brain—in which thought is living, animalistic, primitive, and dangerous.[17] The primal thought is not simply a metaphor for our fears about mechanical access to the mind or connections between thought and intention; it is also a symptom of a theoretical simplification that conflates the mind and the brain, energy and matter. Whereas Merab Ebertle's "The Thought Translator" constructs mind reading as a transparent translation, narratives such as Paul Ernst's "From the Wells of the Brain" (1933) characterize it as a constructive process through which thoughts are born into living matter and develop a volition all their own.

"From the Wells of the Brain" details the demise of Professor Wheeler, a scientist who—akin to Hans Berger—specializes in "neurology and basic electrical energy" and claims that "thought is electrical energy generated by the brain" (1933, 49). After inviting his friend Carson into a darkened laboratory, Wheeler explains his apparatus and experimental hypothesis: "Thought," he argues," is potentially living matter! . . . every human being gives birth to life with every thought he thinks. For his thoughts are living creatures, born in his brain, passing off to invisible lives of their own" (49). On a screen at the far side of the room, Wheeler projects streams of thought particles onto a screen; the particles coalesce into still and moving images: his long-dead brother, his father, an angel, and finally the clearest materialization of all, "a murder thought" that resembles a hairless, lumbering, fanged "gorilla" (52). To create each image, Wheeler uses a helmet apparatus that "increases the voltage of thought approximately a million times" (49) to produce thought projections. Unfortunately for Wheeler, "evil thought materializes with more clarity than good"; during this particular demonstration, the beastly murder thought steps off the screen to terrorize and finally kill its creator (52).

Ernst's narrative, Dr. Wheeler's scientific gaze, and his thought projector are all products and producers of three disciplinary and cultural assumptions about the mind and brain: that we have a hidden/inner self, that we have lower/reptilian brain, and that we have an anatomical object called the brain stem. Dr. Wheeler merges these constructs into a singular object, what he terms the "hidden lower brain" (51). His conglomerate object illustrates the ways science is not simply applied to preexisting problems so much as it helps create problems in need of solutions. Nikolas Rose explains this process in the case of applied psychology, noting that "the social role of psychology should not, therefore be analyzed as a history of 'applications' but as a history of 'problematizations': the kinds of problems to which psychological 'know-how' has come to appear as solutions and, reciprocally, the kinds of issues that psychological ways of seeing and calculating have rendered problematic" (1992, 353). Variations of the hidden self, for example, rendered the unseen and unknown psyche as problematic and in need of reform during the Progressive period, but in psychoanalytic texts of the 1920s the "hidden self" explains and justifies our behavior and motivations, not for reform's sake but as a route to better self-fulfillment (Burnham 1988, 81).[18]

Ernst's narrative plays with the idea of scientific problematizations by creating an object that tests the legitimacy of several disciplinary concepts whose absurdity is only revealed through the cognitive estrangement of science fiction. Dr. Wheeler's form of mind reading, then, is able to reveal several assumptions that inform the literacy of mind reading more generally. Wheeler's work purportedly unveils, for example, "what fills the subconscious *mind* of each of us, just under the thin shell of civilization," "the creeping, hidden thoughts we all have but seldom put into words. A glimpse into the foul, blind, animal *brain* of man!" (51; emphasis added). While introducing the experiments, Wheeler asserts, "The most interesting thing I have discovered, and that which will be of most benefit to science, is the materializing of the more secret and hidden *thoughts*. Things crawling up from the black wells of the hidden lower *brain*. Some of the most beautiful things, Carson, and—some of the most damnable!" (51; emphasis added). Here, the hidden self, the lower brain, and the brain stem are conflated and made to seem indistinguishable from one another, even as each object's history and description mirrors Dr. Wheeler's hidden lower brain in form and content.

The hidden/inner self, which is typically associated with the unconscious, is likewise characterized as animalistic and instinctual.[19] In *Your Hidden Powers* (1923), James Oppenheim describes the inner self as follows.

The unconscious, or inner mind, is not only the source of those things which we call divine. It also has a demonic side. We have inherited from the animals and from the brutal ages of man, terrible things. It is the primitive, the elemental. And when this side of our buried nature is roused, it is like a wild animal come out of his cave, or like an inner thunderstorm or cyclone. (82)

Oppenheim's description draws attention to the assumption that man's deviance and uncivil behaviors are as natural and unstoppable as the weather systems with which they are compared. There is also a sense in which the demonic and the divine are determinable, separable, and clearly associated with the primitive or the civilized. As Freud theorized, and psychoanalytic disciples, such as James Oppenheim, Louis Bisch (*Your Inner Self,* 1922), and David Seabury (*Unmasking Our Minds,* 1924) concurred, the hidden self represents a less evolved moment in man's development, one that is eventually expected to be controlled by the superego. For this reason, these aspects of our personalities only become hidden and invisible under the cover of modern civilization.

Sciences, including neuroscience, would—and would continue to—describe and theorize the lower regions of the brain as similarly primitive and animalistic. The lower brain, also known as the reptilian brain or the brain stem, is said to be the earliest evolved part of the brain. According to a locationalist model, the brain stem controls the automatic functions of the body: reflexes, instincts, and drives—the same physiological processed that lie detection typically monitors. In LeVay's *The Sexual Brain,* published not in the 1920s or 1930s, but in 1993, the hypothalamus is described in strikingly similar terms to Oppenheim's hidden self.

> People tend to stay away from the hypothalamus. Most brain scientists (including myself until recently) prefer the sunny expanses of the cerebral cortex to the dark, claustrophobic regions of the base of the brain. They think of the hypothalamus—though they would never admit this to you—as haunted by animal spirits and the ghosts of primal urges. They suspect that it houses, not the usual shiny hardware of cognition, but some witches' brew of slimy, pulsating neurons adrift in a broth of mind-altering chemicals. (39)

Here, cognition and instinct (a.k.a. reflex) are separated and separable in the brain, yet, even in this description one does not necessarily evolve into or supersede the other. Instead, the civilized and the animalistic coexist much as they do in Ernst's short story.

There is also a hint, in Ernst's, Oppenheim's, and Levay's descriptions, that the hidden self and the lower brain house the potentially dangerous and brutish elements of man—the less evolved behaviors and tendencies that are held in check by the superego. Indeed, the entire development of Wheeler's thought visualizations follow an evolutionary progression: at first "a slight worm of movement appeared, glowing, in the black square's center. . . . At first the thing was merely a quivering speck of luminosity, like a cloud wisp with a light behind it. Then it began to grow. It spread till it filled the screen. And now it was a restless, shimmering sea, with slow-rolling undulations" (50). From this sea eventually comes the form of a man: "The image on the screen ceased rippling. It began to change. It drew up into itself. Shoulders and chest were formed. A face spread into being" (50). The succession of images in which thoughts materialize slowly from a seeming abyss is reminiscent of a colloquial version of evolution: life began as a seething abyss, from which single-celled organisms emerged, by a process that would eventually yield mankind.

And, yet, there is something different encapsulated in the heart of this evolutionary narrative: the being that emerges is not man, but a murder thought. In its form, this living thought is reminiscent of a human "ancestor," the ape; yet in its creation, the murder thought is tied to the future of humanity: a child born not of woman but of science. Our narrator describes the murder thought in these terms.

> It looked a little like a gigantic ape. But it was hairless, with a sickly white pelt that glistened dully, as had glistened the viscous surface of the quagmire. It had arms like thighs, ending in hands thrice as big as ordinary hands; and these hands were half flexed in a throttling gesture, like great talons. The face was slit across by a gash of a mouth in which showed rotting yellow fangs. The nose was simply two pits in the surface of the flat face. The eyes were tiny, bloodshot wells that glittered in the floodlights with the lust to kill. (Ernst 1933, 52)

The ape-man murder thought is representative of primitive man, evolutionarily speaking; but it is simultaneously a child of sorts, the "monstrous creation of Wheeler's brain . . . born of a brain instead of a woman's body" (50). Indeed, for all of his scientific prowess, Wheeler underestimates the power of his invention to actually create living matter, while overestimating his ability to control any thought projection. As the beast begins its menacing march toward the two men, the professor's only response is to "pass his hand before his eyes" and whisper, "it—it isn't possible—It can't have substance—existence!" (53). Yet, even after "everything that had been cunningly concentrated to bring this creature into solid existence had been thrown out of gear, . . . the murder thought had not dematerialized!" (53). Whether Wheeler is directly responsible—because he repeatedly conjured the murder thought—or whether his demise is simply part of a larger commentary on the responsibilities of science, the material consequences remain the same: Wheeler is killed by his own thoughts. His own hidden lower brain is a powerful but perhaps imperfect creator, as was Mary Shelley's Victor Frankenstein.

By combining elements of the hidden self, the lower/reptilian brain, and the brain stem, Dr. Wheeler creates a new object for science to study and repair, namely, the hidden lower brain. This new object not only defines certain types of thought as dangerous but reflects upon the crafting of its components and their own disciplinary and social construction. While the hidden lower brain may seem absurd to us, its constituent parts were—and some still remain—very real disciplinary objects. As we saw with LeVay's description of the brain stem, for example, the neurosciences' popular disciplinary lore has retained some of the fears about the primitive nature of certain parts of the brain; as we shall see in chapter 5, the neurosciences have also continued to elide mind and brain as they problematize and pathologize deception. In the final section of this chapter, I examine one final aspect of mind reading as a lay literacy for brain imaging: "You are what you think," an imagined internalization of the colloquialism "You are what you eat."

### Inside "The Frightening Truth in People"

In each of the theories and narratives I have been analyzing, mechanical mind reading is implicated in the execution of social justice. Secrets are revealed; remembered injustices are illuminated; irresponsible scientists are thwarted. But there were some early indications that this new literacy might also be applied to criminal justice as well. It is no coincidence that "From the Wells of the Brain" is focused on a murder thought, or that in "The Thought Translator" the narrator extols the virtues of a machine that could save mankind: "Through its use the wrong man would never be hanged, or sent to unjust death in the electric chair. Justice could always be meted out" (Ebertle 1930, 6).[20]

The application of mechanical mind reading to criminal justice would necessitate a new way of understanding thought: as evidence for/about a crime and as fodder for criminal reform. We saw an early version of this shift in Hugo Münsterberg's psychophysiological instruments. These mental microscopes were capable, Münsterberg claimed, of making "visible that which remains otherwise invisible" including "lies in the mind of the suspect" (1908, 77–78). Likewise, in "The Thought Stealer" (1930) the connections between mind reading, lie detection, and the courts are explicitly made when the narrator finally traps a criminal scientist with his own machine: "The machine betrayed him. . . . For the first time Collison seemed to realize what his thought-detector would do to him in a court of law" (Bourne 1930, 2). In Alfred Bester's mid-century novel, *The Demolished Man*[21] we have the blueprints for a scientific-telepathic police system that can translate thought energy into not only language and image but also material evidence for criminal charges.

Psychophysiological instruments (such as the sphygmomanometer) that ostensibly read the mind by tracing changes in the body, were deemed inadmissible in U.S. courts (see chapter 1). Indeed, resistance to Münsterberg's technologies specifically, and mechanical mind reading more generally, can be found as early as 1912. Progressive reformers, for example, questioned the reach of science—and its technologies. Thomas Holmes, a member of the Howard Association for penal reform, argued against the broad and indiscriminate application of psychological instruments. In *Psychology and Crime* (1912), Holmes expresses concern about the exposure of the criminal mind, in particular; in his estimation seeing into the criminal mind would not help the causes of punishment, pity, and reform.

> I am still more firmly convinced that it will be a bad day for us when science or evolution provides us with mental rays that will enable us to explore the criminal mind. . . . Punish him if we must! pity him we certainly should! control and reform him if we can! But let us make no attempt to turn him inside out and exhibit his mental organization to curious people by a series of mental photography. (11)

Most important to Holmes is the need to curb the development and use of "mental photography," the "exhibit of mental organization," and the visualization of thought. It would be "better," according to Holmes, "a hundred times better for us to remain in our present state of ignorance, thinking the best of each other, than for us to probe the bowels of unwelcome truth" (12).

Yet, in the decades that follow Holmes's admonishment, numerous lie detectors and tests for concealed information were developed.[22]

Moreover, there was a proliferation of science fiction that predicted criminality would be exposed and cured by the very "rays" that Holmes so adamantly detested but which were being marketed as instruments of social progress.[23] By 1948, J. B. Rhine offered a brief history of and argument for psychical research into telepathy and related phenomena, presenting the sciences of extrasensory perception (ESP) as a remedy for social ills. His concern, particularly after the violence of World War II, was not whether these phenomena exist, but how they could be employed: what effect "they may be expected to have upon relations among men" (166). "Good human relations," he argued in *The Reach of the Mind* (1948), are "not to be based on faith or guesswork" (166); "Our feelings for men depend on our ideas, our knowledge about them" (175). This is all part of a general movement in popular culture and fiction to imagine mind reading as a literacy that could facilitate interpreting the criminal mind in ways that reveal material evidence. And it was believed that such supersurveillance could lead to widespread social reform. Alfred Bester imagines such a world in *The Demolished Man,* a novel that was first serialized in the pages of *Galaxy Magazine* in 1951.

*The Demolished Man* is the story of Ben Reich, a corporate giant in twenty-fourth-century New York City, who tries to get away with murder. While he succeeds in killing his rival, Craye D'Courtney, Reich cannot escape the "Espers" (think ESPers) who, as a kind of psychic police force, telepathically monitor thoughts, emotions, and intentions. Thanks to these "brain peepers," "there hasn't been a successful premeditated murder in 79 years" (Bester 1996, 23). Reich initially outwits telepathic surveillance by creating a "mind block" impenetrable to even 1st-class Espers who are able to "peep" the unconscious levels of the mind.[24] Yet, the guilt welling up in Reich's unconscious is eventually detected by Lincoln Powell, 1st-class Esper, criminological specialist, and lead investigator of the D'Courtney case.

Akin to the other narratives I have explored, the Espers' visualization of thought via telepathy is explained in the novel as a conglomeration of thermodynamics and psychoanalysis. When Espers communicate with each other, they are described as "dualists fencing with complicated electrical circuits" (33). According to Detective Powell, telepaths are not special beings; they have simply developed and/or learned how to capitalize on their energy stores. "Every human being has a psyche composed of latent and capitalized energy. Latent energy is our reserve . . . the untapped natural resources of our mind. . . . Most of us use only a small portion of our latent energy" (235). One faction of the Esper Guild even argues that "telepathic ability [is] not a congenital characteristic, but rather a latent quality of every living organism which could be developed by suitable training" (128).

The thermodynamics of Bester's telepathy are complemented by psychoanalytic conceptions of the mind's drives and desires. The mind reading of 1st-class Espers (able to reach the unconscious) literally and symbolically takes the reader through the various levels of the ego, id, and superego. Yet, the psyche is described as comprised largely of the hidden self and the lower (reptilian) brain. As Powell peeps the character Barbara D'Courtney, who witnessed (and suppressed the memory of) her father's murder, he travels "down the black passages again toward the deep-seated furnace that was within the girl . . . that is within every man . . . the timeless reservoir of psychic energy, reasonless, remorseless, seething with the never-ending search for satisfaction" (148). The description of Barbara's unconscious is reminiscent of Freud's description of the id in *Civilization and Its Discontents* (1929), but it is also akin to the primordial sea figured in Ernst's narrative and the primitive symbolism found in Warcollier's telepathic transmissions. Telepathy's links to primitivism often signal a certain atavism of mind in which the unconscious remains connected to an earlier stage of emotional and communal bond.

Throughout the novel, thought is rendered visible as pictures and through what might be called typographic pictographs. Bester not only describes but illustrates the interlaced nature of telepathic transmissions (fig. 4). Early in the novel, we are invited to an Esper party in which "a weaving, ever-changing, exhilarating design" (33) delineates the conversations. "The wives were arguing violently in sine curves. @kins [Atkins] and West were interlacing cross-conversation in a fascinatingly intricate pattern of sensory images" (35). In some instances, as in "The Thought Translator," patterns are freely formed without attention paid to their construction; this is the case in figure 4. More often, Espers consciously compose their thought patterns (TPs), as is the case in figure 5. Here, a pictographic joke is being broadcast: "What the devil was that? An eye in a glass? Eh? Oh. Not a glass. A stein. Eye in a stein. Einstein. Easy" (34).

In a world monitored by Espers, appearances are of little account, because the lower regions of mind ostensibly reveal the self before it can be masked by any civilizing forces. Although *"people always expect villains to look villainous"* (125), they do not see what Espers understand: that the

Frankly                    Canapes?                    Why
    Ellery,                     Thanks    delicious.    Yes.
        I                        Mary, they're          Tate,
        don't                                            I'm
          think                                       treating
    We          you'll      Canapes?               D'Courtney.
    Brought      be                                     I
    Galen    working                                  expect
    along       for                                    him
        to            Monarch                           in
        help him celebrate. much                          town
            He's        longer.                        very
            just        The                                shortly.
            taken his Guild Exam
          If              is        and
        you're          just        been
    interested          about       classed
        Powell, we're ready              2nd.
                    to
                run rule
                you    Monarch's
                for            espionage
            Guild    Canapes?     unethical.
        President.
                    Canapes?
                    Why yes.
                        Thank
            Canapes?        you,
                        Mary . . .

Fig. 4. Representations of Esper transmissions via textual pictograph from Alfred Bester's *The Demolished Man;* in this image, a thought-conversation

deeper levels of mind determine the self: "The mind is the reality. You are what you think" (28). As we have already seen, the adage "You are what you eat," which imagines your diet and character are mutually imbricated, has been reimagined to include thought as particulate matter; "You are what you think" imagines your thoughts and character to be in a similarly reciprocal relationship.[25] One's thoughts, like one's diet, are expected to be changeable; psychological notions of reform have been cen-

<pre>
The                                                vast,
sea                                                and
is                                    out          Glimmering
calm                         in          the       stand,
tonight,                 tranquil            bay    England
The        Come          to          the window    of
tide       sweet         is          the   night   cliffs
is         air.                          Only       the
full       from                          the        gone;
the        long                       line          is
moon               of spray                         and
lies                                                Gleams
fair                                                light
</pre>

Upon the straights;—on the French coast the

Fig. 5. Representations of Esper transmissions via textual pictograph from Alfred Bester's *The Demolished Man;* in this image, a Rubicon (eye in a stein) for "Einstein"

tered on changing both how and what you think. The main criminal in the novel, Ben Reich, is particularly difficult to identify and apprehend as a criminal, in part because he is so charismatic: his character does not betray his thoughts very readily. "He's got charm," argues Detective Powell, "that's what makes him doubly dangerous," particularly as he manipulates his accomplices (125). But Espers are able to sense beyond appearances. By delving into the furnace, the working energies, of the mind, Espers function as detectors of not only thought, but truth. "We see the truth you cannot see," notes Powell after capturing Reich (243).

The "truth" of which Powell speaks is constructed to be synonymous with knowledge; it is allied to the energetic (that which can be measured) and the visual (that which can be seen). It is also burdensome, in that it encompasses any knowledge one might have about an individual. Espers bear the burden of this knowledge and its prognosis, the diagnostic separation of the normal and the pathological. For this reason, Espers exist in a kind of mental "psychiatric ward. Without escape . . . without refuge," explains Powell to a non-Esper.

Be grateful you're not a peeper. Be grateful that you only see the outward man. Be grateful that you never see the passions, the hatreds,

the jealousies, the malice, the sicknesses . . . Be grateful you rarely see
the frightening truth in people. The world will be a wonderful place
when everyone's a peeper and everyone's adjusted . . . But until then,
be grateful you're blind. (236–37)

For Espers like Powell, "blindness" is a metaphor for tuning out and dis-
connecting from thought energies.

However, "seeing" is not merely metaphorical. Espers' "second
sight"—their ability to *see* the truth in people—is described in visual and
even photographic terms. During his stint as Barbara's peeper and psy-
choanalyst, Powell gathers a plethora of realistic and symbolic images.
He is able to relive and interact with Barbara's memory of the murder
scene, because as Powell argues, "*she's got the detailed picture of the murder
locked up in her hysteria*" (126). Like the memory movies of "The Thought
Translator," Barbara's memories appear to be fixed in terms of content,
but dynamic in terms of access. While it replays over and over, he can
change angles, request her to pause the memory, and even capture images. At one point, Powell cries out "Hold that image. Photograph it";
once outside, he transmits the image of Reich's murder weapon to his
companion Esper, noting, "*Clear picture. Take a look*" (138). This moment
is perhaps most reminiscent of Thomas Holmes's prediction of "mental
photographs" or even Warcollier's suggestion that telepathy works like
"wireless photo." But the implications of an Esper seeing Barbara's memories are far more damning for Ben Reich, whose life and reputation
hinge on this unspoken, yet perfectly reiterated witness's testimony. Who
is Reich? What has he done? Is he a criminal? In no uncertain terms,
Reich is not only *what he thinks,* but what others remember him to be. His
diagnosis depends on the visualizations of his and others' thoughts and
memories.

As they code and translate thought, Espers are figured as the latest
machines whose mind reading abilities surpass their aged ancestor: the
Mosaic Multiplex Prosecution Computer (a.k.a. Mose). "Old Man Mose"
is figured as "that mechanical brain" (171)—an anthropomorphic combination of circuitry whose "multitudinous eyes winked and glared
coldly. His multitudinous memories whirred and hummed. His mouth,
the cone of a speaker, hung open in a kind of astonishment at human
stupidity. His hands, the keys to a multiflex typewriter, poised over a roll
of tape, ready to hammer out logic" (169). He is the consummate vestige
of logic, rationality, and objectivity; however, Mose is also an "old man"—
a somewhat ailing and outdated version of judicial logic who is unable to

gather evidence; he works from a "multitudinous" but fixed memory and "hammers" out a singular logic, without attending to the ambiguity and ingenuity of human thought. While Mose is able to solve the case, "Old Man Mose was right" (191), he does so only after Powell has performed the brain-work.

Indeed, Powell's telepathic abilities are figured in equally mechanistic terms. As Powell prepares to search for Reich in the jungle Reservation, for example, he considers one logistical problem: How can he locate Reich in twenty-five hundred square miles of terrain if there are "no mechanical devices outside of cameras allowed on the Reservation" (160)? His answer: "I'm going to do some fast co-opting and take my own Radar into the Reservation." His "radar" is, of course, peeper-powered. To ensnare Reich, he enlists the help of hundreds of Espers who converge on the Reservation. Together, they create a telepathic dragnet in which "the TP Band crackled as comments and information swept up and down the line of living radar in which Powell occupied a central position" (162).

However, for all of their access to the thought, truth, and knowledge triumvirate, Esper knowledge is not considered objective evidence. Like Marston's systolic blood pressure test for the symptoms of deception, Esper knowledge "isn't admitted in court" (23); it must be filtered through established systems of knowledge production, in this case, MOSE, the Mosaic Multiplex Prosecution Computer who requires "motive, method and opportunity . . . and insists on hard fact evidence" (87). In this respect, Espers are the middlemen, the translators that process thought energy to produce information that can be coded and entered into Mose. Powell's trips into Barbara's unconscious provide a telling example. As Hugo Münsterberg speculated in *On the Witness Stand*,[26] witness testimony is the product of Barbara's memory and therefore subject to her own unconscious processes, as were the thought visualizations in Stanley Weinbaum's "The Ideal" and Paul Ernst's "From the Wells of the Brain." Barbara's memories are rendered particularly questionable and subjective as we view the deeper portions of her subconscious in which Powell sees himself as a father/lover figure. Connections to Powell as paternal return Barbara to images of her biological father, D'Courtney, in what amounts to a matrix of latent associations.

> The images of Powell-Powerful-Protective-Paternal rushed at him, torrentially destructive. He stayed with it, grappling. The back of the head was D'Courtney's face. He followed the Janus image down to a

blazing channel of doubles, pairs, linkages, and duplicates to—Reich?
Imposs—Yes, Ben Reich and the caricature of Barbara, linked side to
side like Siamese twins, brother and sister from the waist upward.
(150)

Somewhere in her subconscious, Barbara realized that Ben is her half
brother—the illegitimate son of D'Courtney himself. In this case, insider
information (pun intended) provides a lead for Powell's investigation;
however, Barbara's subconscious is not code-able evidence as of yet. After emerging from Barbara's psyche, he explains the process to a fellow
Esper as follows: "Trying to make sense out of fragments in the Id is like
trying to run a qualitative analysis in the middle of the sun. . . . You aren't
working with unit elements. You're working with ionized particles"
(151). Once again, we are reminded that thought is energy, and energy,
matter; yet here, even particulate matter does not inspire the kind of
surety it does in other narratives. This difficulty with the imagined particulate matter of thought helps to cognitively estrange us from other,
similar assumptions about the body and character. Neither physical bodies nor particulate thoughts are the best matter for evidential debates.
    Moreover, Powell has several flaws that hinder his so-called objectivity
and call into question the state's reliance on mind reading as a literacy
for understanding "the truth in people." Ironically in a novel about the
detection of deception, the detective is best known for his dishonesty.
"The trouble with Powell," in particular,

> was an enlarged sense of humor, and his response was invariably ex-
> aggerated. He had attacks of what he called "Dishonest Abe" moods.
> Someone would ask Lincoln Powell an innocent question, and Dis-
> honest Abe would answer. His fervent imagination would cook up the
> wildest tall-story and he would deliver it with straight-faced sincerity.
> He could not suppress the liar in him. (27)

It is in this very humanity, this ability to creatively deceive, that Powell's
powers of observation trump Mose. When it comes to the final negotiation of capturing Reich and saving the life of his hostage, Hassop, Powell
insists that they leave the Reservation. "*This needs finesse,*" Powell telepathically transmits, "*I don't want Reich to know I'm abducting Hassop. It's all
got to look logical and natural and unimpeachable. It's a swindle*" (164). Powell's comment illustrates the utility of logic, not as judicial proof or unimpeachable evidence, but as a cover—as a means to disguise deception.

The construction of thought as evidence, then, reflects not only on our conceptions of thought but on our understanding of good evidence and evidence-gathering procedures. We already saw arguments, in chapters 1 and 2, that would end third-degree interrogations; here, mechanical mind reading is offered as one more solution—perhaps even a better way to get a confession out of a suspect. However, in Bester's novel, thought is not always easy to translate: it is particulate, sometimes uncooperative, volatile, and symbolic. In addition, the very detectives who are entrusted with this literacy are neither held in check nor held accountable for their own deceptive techniques. In its affiliation with thought, evidence is reimagined as not fact but knowledge, and skewed knowledge at best.

While Thomas Holmes's mind ray, Hugo Münsterberg's mental microscope, and Alfred Bester's Esper police force have not come to fruition, their metaphorical conceptions of mind reading machinery reveal much about a lay audience's hopes for and fears about the visualization of thought for the criminal justice system. As in the advertisement with which I began, the desire is to "get mind reading down to a science" that can then be applied to any number of situations. Among other troubling aspects of this particular narrative, thought as evidence raises some very difficult issues in terms of the Fifth Amendment, for example, that neuroethicists are still attempting to untangle. A cultural history of mind reading can help as we grapple with these conundrums, by explaining where and how the lay public and the media came to believe in these literacy narratives.

The visualization of thought in science and science fiction between the 1930s and the 1950s can tell us much about the visualization of the mind we now regularly see in popular representations of brain imaging technologies. As I noted at the outset, critics have long traced this conflation of mind and brain; surprisingly, what they have not examined is the history of this elision in the popular lay texts of science fiction. By tracing the cultural history of popular narratives associated with the literacy of mind reading, we can better understand media representations of brain imaging like the one with which I began. Instead of dismissing advertisements and newspaper copy as nonscientific or hyperbolic, we can appreciate them as a valuable component in the cultural history of brain imaging technologies.

While this genealogy has many strands, some others of which I trace in this book, I have explained four ideological assumptions central to the literacy of mechanical mind reading: that thought is energy; that thought energy is transparently translatable and related to character; that primitive and dangerous thoughts have a definitive location; and that thought could be used as material evidence. As a group, these ideological assumptions help to explain why, at least in the popular imagination, mechanical mind reading seems plausible: thought is matter, and matter can be quantified, monitored, and controlled; particulate thought, like the physical body, can be understood as a manifestation of character.

The next chapter takes up the issue of patterns written on the body and the brain, and their systemic application to individualization and classification, through an analysis of the history of fingerprinting via *The Invasion of the Body Snatchers*. What may at first sound like a digression actually returns us to the heart of the issues I have been discussing in this chapter and will continue to analyze in chapter 5: that the body—particularly the deceptive body—has been constructed in and through assumptions about the brain and the mind; and that these assumptions have been taken up by not only psychology but also neuroscience as a means of diagnosis, evaluation, and intervention.

# 4

## *Without a Trace*

### Brain Fingerprinting, Biometrics, and Body Snatching

"Not only your brain, but your entire body, every cell of it emanates waves as individual as fingerprints. Do you believe that, Doctor?"

—JACK FINNEY, *Invasion of the Body Snatchers* (1955)

From Luther Trant's examination of the marks of crime on men, to William Marston's hypotheses about the deceptive consciousness, to the mechanics of mind reading, I have illustrated several ways in which literature, science, and the media share the assumption that deception leaves a trace in our physiology. Instead of merely looking to the crime scene for fingerprints and bloodstains, police have been asked to turn their gaze on the suspect in new and technologically mediated ways. Indeed, the body's internal physiological mechanisms are central to both historical and contemporary lie detection: blood pressure, pulse, and respiration are all said to increase as a result of prevarication, as are blood oxygenation levels and brain waves.

That we look to the body to identify deception is certainly problematic, for the reasons I have been outlining. However, this body-centricity is congruent with many biometric techniques, including Alphonse Bertillon's nineteenth-century anthropometry, and its eventual rival, Francis Galton's fingerprint classification system. In recent memory, the sequence of our DNA and the shape of our retinas, among other measurements, have joined fingerprinting as anatomical markers of uniqueness, with DNA now surpassing fingerprinting as the best method for identification of criminals and civilians alike.

Yet, among biometric technologies, fingerprinting has achieved, both practically and metaphorically, the most pervasive influence on how we understand the body as a self-reporting entity that will divulge information about identity and individuality—but not character[1]—while thwarting attempts at deception. Early systems of fingerprint identification, like many biometric systems of identification employed today, depended on the presumption that one's unique identity can be traced through the measurement of various aspects of one's anatomy: that biology is intimately connected to an individual's identity. Or, put another way, "fingerprinting has embedded firmly within our culture the notion that personhood is biological. The idea that our individuality is vouched for by our biological uniqueness" (Cole 2001, 5). It is no wonder that new biometric technologies often appropriate the word *fingerprinting* and use it in their names: DNA (or genetic) fingerprinting was coined in the 1980s to explain the process of testing cells for unique, identifying markers; by the 1990s, Brain Fingerprinting emerged as a technology that could ostensibly provide a unique catalog of the "information stored in the human brain" (Farwell 2001a). Both techniques assume that individuality is "vouched for" in our anatomy, and both rely on a metaphoric connection to fingerprinting as a way to ground their basic claims about the absolute and static nature of the body's individual identity.

In chapter 5, I analyze the discursive presentations of Brain Fingerprinting and fMRI in scientific and journalistic publications; here, I make two preliminary points: first, classification systems are not the revelation of nature's order;[2] instead, classification and its objects are co-constitutive. Take, for example, Londa Schiebinger's extended analysis of Linnaeus's taxonomical system, which, she argues was the result of sociopolitical ideology rather than "objective" discoveries about nature's inherent organization (1993). I place "objective" in scare quotes given that we have already encountered the genealogy of this term in chapter 2 and because certain definitions of objectivity have been instrumental in defining classificatory systems. These definitions, which have taken several forms over the past three decades, are best encapsulated by what Sandra Harding has called "weak objectivity," what scholars such as Robert Proctor have termed "value free science," and what Donna Haraway has theorized as the "god-trick": the belief that objectivity is equivalent to an uninvested view from nowhere. For each of these theorists, a more responsible definition of objectivity includes the admission of bias, position, situation, and context.

Feminist theorists Karen Barad and Donna Haraway have insisted that responsible conceptualizations of objectivity will also account for the interplay of the material and immaterial, the constructed and the real. As an interlocutor for Judith Butler, Karen Barad argues specifically that "objectivity means being accountable to marks on bodies" (Barad 2003, 824). In the context of fingerprinting's history, we could interpret Barad's statement thusly: objectivity does not mean merely counting, noting, or recording bodily stigmata, such as fingerprints. Accounting for marks on bodies entails accounting for the scientific and cultural systems of classification that bring *those particular marks* to the fore and employ them for the classification of bodies. In the case of fingerprinting, Francis Galton and J. Edgar Hoover, among others, constructed the fingerprint as a product and producer of the self-reporting body.

A related, secondary argument in this chapter is that cultural representations, such as Jack Finney's *Invasion of the Body Snatchers,* literally and figuratively explore the internalization of individuation by suggesting that the body is a fulcrum for classificatory knowledge production *and its undoing*. In the novel's fictional Santa Mira, California, originals and duplicates cannot be distinguished by traditional systems of identification; indeed, the most deceptive bodies are unremarkable not because of but despite their marked status. Yet, even as individuality is linked to and inscribed in the biometrics of bodies—through fingerprints and cellular blueprints—identity also hinges on affective responses whose meaning exceeds the limits of our classificatory systems even as it is circumscribed by them.

## The Body's Self-Signatures

The development of Western fingerprinting systems[3] took place at a time when the British government sought to intentionally mark foreigners (and only later, criminals) as physically distinct, as measurably different from one another.[4] Indeed, finger- and palm prints were first used to classify colonial populations by Sir William Herschel who employed them to identify pensioners seeking assistance from British government outposts in India in 1858 (Browne and Brock 1953, 33). Not surprisingly, his system of "personal identity" was not without political motivation. In fact, Herschel began to rely on fingerprints "immediately after the eruption of the Indian Mutiny [1858] against British rule" (Thomas 1999, 217), implying that one goal was to reestablish waning British colonial control. Sir Francis Galton's famous text, *Finger Prints* (1892), which helped to solidify the use-value and application of the technique for British and American criminalistics, is often cited as the foundation of fingerprint identification systems;[5] however, Herschel's experiments in the colonies preceded Sir Francis Galton's monograph by nearly forty years. And long before Herschel, individuals in many countries, including China and India, used palm and even foot marks as signatures for contractual agreements (Lee and Gaensslen 1991).

Yet, as the British began to use finger *prints*—instead of marks—for documentation, they also assumed possession of the technique. Herschel's retrospective narrative history of fingerprinting, *The Origins of Finger-Printing* (1916), denies the legitimacy of digital systems[6] of identification used in other parts of the world,[7] while reaffirming the "experimental" nature of his investigations in India. He does so explicitly in an appendix to the tome that covers any and all references to digital marks preceding his own work and that of Galton, a list that includes Thomas Bewick, Johannes Purkinje, and indigenous peoples in Bengal and China. Each historical use of the finger's mark is deemed helpful but incomplete. When commenting on the Chinese use of digital marks, for example, Herschel tells us that "the science of identification by means of the pads cannot, in my opinion, date further back than 1858, when I happened to use oil-ink, which was not used for *tep-sais* [a finger-tip ink blot]" (40). Later, he explains in no uncertain terms that the main difference between the British system and all others is that "these marks were not made, as ours are, expressly to challenge comparison; that, in fact, they offer no points for comparison" (41). "In conclusion," Herschel skeptically writes, "it is hard to believe that a system so practically useful as this could have been known in the great lands of the East for generations past, without arresting the notice of Western statesmen, merchants, travelers, and students. Yet the knowledge never reached us" (41). With this final line, Herschel reaffirms not only the power, dynamic grasp, and omniscience of British national identity, but also the right to control and commodify the knowledge of other cultures/colonies.

The prints Herschel collects are granted far more currency, literally and figuratively, than the marks or smudges of other cultures. His book contains the hand prints of an Indian merchant and a facsimile of a Chinese bank note stamped with a fingerprint that Herschel categorizes as a mere "smudge" (39). Their inclusion in the British historical record performs the marking—even specimenization—of the Other. Toward the beginning of *Origins of Finger-Printing,* Herschel argues that the British "*possession* of [a fingerprinting system] derives from the impression of Konai's hand in 1858" (1916, 9; emphasis added). The event to which he refers is an encounter with a contractor named Konai. Instead of asking the man to sign his name, Herschel required him to ink his entire hand and place a print on the back of the document. After their transaction, the hand print is given as a gift to Sir Francis Galton who "presented" it as an artifact to numerous other Englishmen at the Royal Society. Herschel joyfully admits, "I was so pleased with the experiment that, having to make a second contract with Konai, I made him attest it in the same way. One of these contracts I gave away to Sir Francis (then Mr.) Galton for his celebrated paper read before the Royal Society, November 1890, to which body he presented it" (9).

Fingerprints offered a solution to the colonial dilemma of visual identification by removing any need to examine or distinguish between faces, instead reading identity through marks on the fingertip, placing them within a system of classification that constructs and concretizes identity. Herschel's use of fingerprinting in his relations with indigenous peoples demonstrates how fingerprinting helped to mark undistinguished subject populations and reified presumptions that the "lower races" were always already suspect. It should be noted that only Konai—not Herschel—was required to stamp his print. Herschel attempts to control the power dynamics by enlisting the authority of a scientific system, one that is reminiscent of the lie detector in its ability to encourage certain "truthful" behaviors: he "made" Konai sign the contract with a physically identifying mark. Herschel's purpose, as he describes it, was "to frighten Konai out of all thought of repudiating his signature hereafter" (8).

Later, the same ideologies of control and assumptions about the behavioral characteristics of certain subject populations inform and are augmented by Francis Galton's systematization of fingerprinting. "In civilised lands," notes Galton in *Finger Prints* (1892), "honest citizens rarely need additional means of identification to their signatures, their photographs, and to personal introductions" (148). By speaking of "civilised lands" and "honest citizens" Galton demarcates the division between colonized foreigners and their European colonial counterparts. The "civilised" are implicitly identifiable, comprehendible, "honest," and therefore do not need to verify their citizenship or personal character through anatomically based (and potentially "essentialized") systems of power. Yet, "in India and in many of our Colonies the absence of satisfactory means for identifying persons of other races is seriously felt. The natives are mostly unable to sign; their features are not readily distinguished by Europeans; and in too many cases they are characterized by a strange amount of litigiousness, willness, and unveracity" (149). Already in Galton's 1892 text, processes of identification involve the finger, not the face, whose "features" are "not readily distinguished by Europeans."

Galton's fingerprint system was in direct competition with an older biometric identification system, Bertillonage, that was developed in 1882 by Alphonse Bertillon, a French police officer. Bertillon's system was made up of three elements, a *portrait parlé* (a detailed description of the person), eleven anthropomorphic measurements that ranged from height to the length of the left middle finger, and photographs that look much like contemporary mug shots. All of this information was recorded on anthropometry cards and was thought to individuate criminals whose measurements had been correctly gathered. Bertillon's system faced several problems, including the fact that his measurements were difficult to collect. Nevertheless, Bertillonage was adopted in Europe and the United States by criminologists who desired to differentiate between members of the criminal population.

Aware of his competition, Francis Galton used *Finger Prints* as a space to craft careful rhetoric about the benefits of his new identification system. Specifically, he argues for "two great and peculiar merits of finger prints; they are self-signatures, free from all possibility of faults in observation or of clerical error; and they apply throughout life" (1892, 168). Touted as the unchanging "self-signature" of the body, fingerprints become products and producers of the trend toward representational objectivity I discussed in chapters 2 and 3. Through technical and/or classificatory intervention, the body appears to divulge information, to speak for itself. And yet, as with Marston's significant lying curve, there is much mechanical intervention between the body and its "self-signature." As Galton outlines in his text, fingerprints must be collected using particular methods, analyzed using particular tools (a magnifying glass, for example), and carefully indexed using standardized types (whorls, loops, and arches).

All of this is not to say that fingerprints—a.k.a. our papillary ridges—cannot exist outside of this system of collection and classification; it is to say that their merits and uses are rhetorically constructed so that they become useful in larger narratives about the body's ability to speak for it-self. Throughout *Finger Prints,* for example, Galton builds upon the idea of self-signatures to argue that fingerprints are less subjective than Bertillonage; in his identification system, "there is no prejudice to be overcome in procuring these most trustworthy sign-manuals, no vanity to be pacified, no untruths to be guarded against" (2). Unlike Bertillon an-thropometric systems,[8] which require multiple measurements of various body parts that may change shape or size throughout life, fingerprints are expected to remain the same, and it takes little skill to collect them from subjects and suspects alike.[9] Galton's argument is bolstered and ex-tended by an excerpt from M. Herbette, the director of the Penitentiary Department in France, who argues that "to fix the human personality, to give to each human being an identity, an individuality that can be de-pended upon with certainty, lasting, unchangeable, always recognisable and easily adduced, this appears to be in the largest sense the aim of the new method" (Galton 1892, 169). Having experimented with matrices of heredity and criminality, composite criminal photographs and eugen-ics, Galton's argument did agree with much of Cesare Lombroso's and Havelock Ellis's criminological work.[10] However, his fingerprinting sys-tem attempted not to define a criminal type but a criminal individual (Thomas 1999, 211). As we shall see, it is this concept of identity, this ob-jective science of bodily marking and classification that *Invasion of the Body Snatchers* problematizes.

As is somewhat evident from the inclusion of M. Herbette's excerpt, *Finger Prints* marked the shift from fingerprinting foreigners to finger-printing criminals[11;] by the 1890s fingerprints were being used regularly in criminal investigations;[12] by 1901, the Criminal Record Office was es-tablished at Scotland Yard; and in 1905 fingerprints were accepted as courtroom evidence in an English murder trial (Lee and Gaensslen 1991, 29). U.S. police adopted fingerprinting in 1904 in a cooperative effort with Scotland Yard officials to safeguard the queen's jewels at the St. Louis State Fair.

## Fetishizing the Fingerprint

Between 1930 and 1960—the same era in which *Invasion of the Body Snatchers* was written, published twice, and made into a film—finger-printing was employed not only as a forensic technique but also in its his-torical guise as means of social control—this time in an American context. Thanks in large part to J. Edgar Hoover, then director of the FBI, fingerprinting became the standard for criminal and noncriminal identification. In so doing, Hoover cemented the links between xenophobia and criminality under the banner of systemic surveillance and biometric identification in the United States. This meant that ideologies of marked alien bodies continued to haunt the civil application of the technology. "After a brief period in which fingerprinting was restricted to people convicted of crimes," notes Simon Cole, J. Edgar Hoover "returned fingerprinting to its origins, as a mechanism for state monitoring and surveillance of citizens, especially those deemed foreign, politically radical, or otherwise dangerous" (2001, 247). Indeed, "the American deployment of this technology on a national scale coincided with a period of particularly intense anxiety about criminal conspiracies, economic instability, and, perhaps most importantly, dangers posed by foreign influences within the nation" (Thomas 1999, 257).

Hoover "made a fetish of both fingerprints and the crime lab" (Walker 1977, 73) in several stages. During the 1930s, Hoover convinced Congress to cede control of state fingerprint records to the Bureau of Investigation.[13] By 1932, he had established the first national crime lab to use these records. He then renamed the Bureau the Federal Bureau of Investigation in 1934. Finally, Hoover launched two movements connected by fingerprinting: a police professionalization campaign in the 1930s and 1940s and an anticommunist crusade in the 1950s. In 1935, Hoover and August Vollmer organized a universal fingerprint campaign through which all citizens were encouraged to file their prints with the FBI.[14] Among the more famous participants were John D. Rockefeller Jr., Walt Disney, and President Franklin D. Roosevelt.

According to the rhetoric of the 1930s campaign, criminals are characterized not only as indistinguishable from the average citizen but as masters of deceit capable of manipulating and thwarting the decentralized penal system. "The day of the mask and the dark lantern is over. Crime lives next door to you. Crime often plays bridge with you. Crime dances with your sons and daughters. It is ever present. But this veneer of social grace that our criminals have adopted in no way makes them less foul" (Hoover 1936c). Here, crime is personified as having fully infiltrated the domestic sphere, as a singular figure who is indistinguishable from the good, normal citizen.

Hoover's solution to Crime was a centralized clearinghouse, the newly renamed FBI, that was equipped with the newest scientific techniques. These techniques, including fingerprinting, promised to unify and make uniform the government's ability to identify and punish those who broke the law.[15] In conjunction with this criminal campaign, a civilian fingerprint registry was developed that promised to protect the law-abiding citizen. Anyone who chose not to submit their fingerprints would be "failing to avail himself of a bulwark against the imposter, the schemer, the faker and grafter who may at any time bring him annoyance, loss of money, and distinctly unfavorable publicity" (Hoover 1936d, 7).

The 1950s campaign followed the second Red Scare of Communist invasion. Hoover, who published *Masters of Deceit: The Story of Communism in America and How to Fight It* (1956), instituted radical forms of governmental surveillance best embodied in figures such as Joseph McCarthy who helped establish a particular status quo. As literary historian and analyst Katrina Mann notes, "anxieties about communism, anticommunism, radiation, and the assimilative impulses of an emergent technocracy were frequently posed as threats to white, patriarchal, or heterosexual . . . primacy"; indeed the "atmosphere of postwar American anticommunism was used domestically as a discursive and political tool to protect the sociopolitical status quo" (2004, 51). In the case of both the 1930s and 1950s campaigns, Hoover's powerful position allowed him to employ forensic techniques at a civilian level in the name of national security.[16]

By the time Jack Finney's "The Body Snatchers" was serialized (1954) and collected as a novel (1955) the FBI still reigned as the best organization to help free America from an "alien" invasion. Critics have argued that the FBI—and, by extension, Hoover himself—is the only governmental institution in Finney's novel to remain unaffected by the invasion of the body snatchers. "Indeed, of the American institutions mentioned in the novel, ranging from Eisenhower to the telephone company, only the FBI remains sacrosanct: it alone is never doubted" (Johnson 1979, 12). In fact, the initial *Colliers* serialization of the novel relies on the FBI as a deus ex machina that arrives just in time to provide reinforcements in the final scene.[17]

Given that the novel makes specific reference to two cultural contexts—Hoover's FBI in the 1950s and the historical construction of fingerprinting as a means to codify and solidify individuality—it should be no surprise that the protagonists' first encounter with the alien Other centers around a set of blank fingerprints. As we shall see in the next sections, the pod-people are literally unmarked before they are fully formed and unremarkable after receiving their final stamp of individuality: a set of fingerprints that are identical to the body they have copied.

## Deceptively Human

Set in 1953,[18] Jack Finney's novel[19] *Invasion of the Body Snatchers*[20] (1955) represents a cultural moment during which identities—particularly the identity of the alien, the foreigner, and the communist—were being reconstructed as crucial to national security.[21] Critics have long associated the novel with Cold War anxieties, including conformity, communism, and the science of the atomic bomb (Mann 2004; Seed 1999; Hoberman 1994; Johnson 1979); in so doing, they often draw connections between the pod-people and collectives whose unseen infiltration could potentially succeed before being unveiled: from communists to McCarthyites, on the one hand, and immigrants and minorities, on the other (Mann 2004). While *Invasion of the Body Snatchers* expresses fears about alien invasion and racial integration, particularly in terms of their effects on suburban American values, the novel is not just an allegory about assimilation gone awry.[22]

What critics have overlooked thus far is the novel's investment in and reliance on various methods of identification and locations of individuation, including the fingerprint. As object and trope, the fingerprint is comparable to what Bernadette Wegenstein has termed "faciality" (2002). Once the locus of identity or a "window into the soul" (Brumberg 1997, 62), the face has lost purchase since the late nineteenth century, a process evident in the shift from Bertillonage to fingerprinting I have been detailing. In "Getting Under the Skin, or, How Faces Have Become Obsolete," Wegenstein argues that over by the past century, the face has also been revealed as a "code" whose "role can be taken over by any other body part" (2002, 235). This code of identity and individuality—even of the soul—has shifted inward: "the priority of the face—in other words, faciality—has moved into the body, to organs, DNA, and other important hidden 'information'" (234). In using the term *code,* I refer both to Wegenstein's definition and to Simon Cole's distinction between *marker* and *code* (2001, 309). In both cases, *code* is more than descriptive, it is prescriptive: a set of "blueprints" (310) that give meaning to a marker.

In *Invasion of the Body Snatchers,* the face quickly proves inadequate for identification; the fingerprint (as object and trope), which takes its place and moves ever inward (to cellular blueprints and even brain waves), also proves inadequate. In these moments, identity is figured as excessive to classificatory systems. By paying attention to literal and figurative representations of fingerprinting in the novel, we can examine the ways that fingerprinting becomes a code for individuation that has been imperfectly internalized at the cellular, neurological, and psychological levels. At the same time, we can question the novel's attempt to reinscribe any identificatory excess as an affective trace. Both strands of inquiry help to complicate the relationship between body and emotion I have been theorizing throughout the book.

Finney's novel begins with Santa Mira, California, being invaded by aliens: pods that take the form of whatever organic substance will help them survive on a host planet. On Earth, "the pods arrived, drifting onto our planet as they have onto others, and they performed, and are now performing their simple and natural function—which is to survive on this planet. And they do so by exercising their evolved ability to adapt and take over and duplicate cell for cell, the life this planet is suited for" (1996, 174). The pods survive on Earth by stealing and replicating the biological blueprints of humans—blueprints explicitly compared to fingerprints and akin to DNA—in a process that causes the original body to disintegrate. The resulting pod-people look like humans, act like humans, sound like humans; indeed they slowly infiltrate the town and adopt its inhabitants' identities right down to the scars on the backs of their knees. Yet, a cast of four heroes (including an author and a doctor) recognize that there are differences between the originals and the replications, differences that cannot be marked, measured, or otherwise quantitatively classified, even by fingerprinting, the most widely accepted system of biometric identification in use in the 1950s.

The first and most notable appearance of fingerprints and fingerprinting in *Invasion of the Body Snatchers* coincides with the initial encounter with a pod-person. After recounting a few days of strange behavior on the part of his patients and friends, Dr. Miles Bennell must confront one of the half-formed pod-people in his friends' basement. Splayed out on the pool table, the lifeless body is not exactly dead—no rigor mortis has set in, no wounds are visible—but it is not yet alive, either. The body appears to be a replication in progress: about the same height, weight, and stature as Miles's friend, author Jack Belicec. As the two men slowly and independently come to this conclusion, Miles has an idea. He grabs some stationery, some ink, and Jack's hand: "I took it, then pressed the fingers, carefully rolling each one on the sheet of paper, getting a full set of clean, sharp prints" (Finney 1996, 42). Prints in hand, Miles retreats to the basement (and the lifeless body) to perform the same operation, a scene which I reprint here in full.

> I picked up the right wrist, concentrating on that, not looking at the face. I inked the ends of all five fingers, then I laid the sheet of paper containing Jack's fingerprints on the wide table ledge, beside the body's right hand. I brought the hand up, laid it on the white sheet, and rolling each finger, I took an impression of them all, directly under Jack's prints, then lifted the hand from the paper. . . . Becky actually moaned when she saw the prints, and I think we all felt sick. Because it's one thing to speculate about a body that's never been alive, a blank. But it's something very different, something that touches whatever is primitive deep in your brain, to have that speculation proved. There were no prints; there were five absolutely smooth, solidly black circles. (43)

That Miles looks to the fingers and not the face is, arguably, not an intuitive assumption, but one related to the prominence of fingerprinting identification and the internalization of the code of individuation once ascribed to "faciality" (Wegenstein 2002, 234). Like other bodies that have unidentifiable faces, this body *should* still carry an imprint of its individuality. Miles and his fellow investigators feel "sick," at least in part, because they are confronted with a blank body, one that is unmarked by any self-signatures. Miles reminds us of his queasiness by returning, on four separate occasions, to the "impossibility" of a body without prints.[23] And yet, the pod-people do not have fingerprints from "birth"; they develop them later, along with other markers of lived experience, including scars, wrinkles, and moles. Although the end product remains the same (i.e., a body marked with signifiers that have been used for classification) the pod-people's acquisition of markings defies assumptions about the body as self-reporting and individual.

If the blank vagueness of the body on the pool table disturbs Miles, Becky, and the Belicecs during their first encounter with the pod-people,[24] what disturbs the protagonists about their later encounters with pod-people is the exactness of their anatomical replications. Indeed, the particular irony of the novel is that the marks that should distinguish individuals are empty signifiers—instead of Galton's self-signatures of the body that promise to reveal, they are the very features that promise to deceive. When the alien pods slowly gain layers of physical characteristics in the process of replication, they remain paradoxically unremarkable not despite but *because* they eventually replicate the bodily markings and patterns—fingerprints, scars, moles, and wrinkles—of their originals. When Miles finds a would-be pod-double for Becky, he catalogs each of the anatomical marks that should distinguish Becky from other humans. Instead, these same marks would complete the perfectly deceptive pod-body.

> There was a scar on the left forearm of the thing on the shelf, just above the wrist. Becky had a small smooth burn mark there, and I remembered its shape because it crudely resembled an outline drawing of the South American continent. It was on this wrist, too, barely visible, but there, and precisely the same in shape. There was a mole on the left hip and a pencil-line white scar just below the right kneecap; and although I didn't know it of my own knowledge, I was certain that Becky, too, was marked in these very same ways. (61)

Miles's final certainty "that Becky, too, was marked in these very same ways" does not depend on his "own knowledge" of Becky's body but on his knowledge of anatomical individuation that the pod duplication process has perfected. In short, and as Becky explains, "Miles, there *is* no difference you can actually see" (17); or, as Miles later concludes, echoing J. Edgar Hoover's sentiments, "They were each our enemies, including those with the eyes, faces, gestures, and walks of old friends" (167–68).

Instead of providing individuation, fingerprints—as object and, especially, as code—become the measure of successful superficial, cellular, and neurological replication. As with Wegenstein's theories of "faciality," the fingerprint leaves its figurative/coded mark of uniqueness on other, internalized substrates of individuation. Explicitly, fingerprints are used as a comparative for cellular energy waves; implicitly, fingerprinting informs discussions of DNA-like cellular replication processes. In each case, fingerprints are marshaled as an authoritative, material-discursive foundation for an individuality that marks the body on the inside as well as the outside.

The processes of duplication are explained in depth by two men whose professions and perspectives are not inconsequential: L. Bernard Budlong, a biology professor; and Mannie Kaufman, a psychologist. Professor Budlong's exposition of pod replication rhetorically connects physiological individuation with ideologies of fingerprinting. Budlong, a botanist turned pod-person, explains that humans are marked not only by their fingerprints but by *unique patterns* at the cellular level. "Your body contains a pattern, all living matter does—it is the very foundation of cellular life. . . . it is a pattern—infinitely more perfect and detailed than any blueprint could be—of the precise atomic constitution of your body at exactly that moment, altering with every breath you take" (176). Budlong's explanation is reminiscent of the intricate *patterns* used to distinguish fingerprints—or the coining of the term *cellular blueprint*. Importantly, these cellular patterns are explained in terms of the internalization of another, older, biometric technology, the fingerprint, by Budlong himself: "Not only your brain, but your entire body, every cell of it emanates waves as individual as fingerprints" (175). Body snatching works by stealing this pattern and applying it to a new set of cells.

In its reference to replication and reproduction, Budlong's explanation is also reminiscent of the influence of fingerprinting on theories about DNA's uniqueness. Indeed, the discovery of DNA implicitly informs *Invasion of the Body Snatchers,* given the timing of its "discovery" in 1953, just a year before the first serial was published.[25] Knowledge of nucleic acids dates back to at least 1869,[26] but Watson and Crick enhanced our knowledge of the configuration, function, and individuating properties of DNA with their publication of "Molecular Structure of Nucleic Acids—A Structure for Deoxyribose Nucleic Acid" in a 1953 issue of *Nature.*[27] Of particular interest is DNA's ability to transmit physiological information from one body to another through what Watson and Crick call "a possible copying mechanism for the genetic material" (Watson 1953, 737). So, although DNA is not mentioned by name in Finney's text, it arguably informs the descriptions of alien replication. Take, for comparison's sake, the *New York Times'* uptake of Watson and Crick's discovery of DNA: "In all life cells, including those of man, DNA is the substance that transmits inherited characteristics such as eye color, nose shape, and certain types of blood and diseases. The transmission occurs at the vital moment of mitosis or cell division when a tangle of DNA containing chromosomes becomes thicker and the cell separates into two daughter cells" ("Clue" 1953, 17). The vital moment described in the *New York Times* is the very moment of which the pod-people take advantage: "The intricate pattern of electrical force-lines that knit together every atom of your body to form and constitute every last cell of it—can be slowly transferred. . . . And what happens to the original? The atoms that formerly composed you are—static now, nothing, a pile of gray fluff" (Finney 1996, 176). The pods replicate their human hosts by duplicating these physiological patterns; the individuality of these patterns allows the pod-people to replicate fingerprints, scars, and behavior on and through the duplicate bodies.

And yet, despite Budlong's explanation of energetic "electrical force-lines," replica bodies ultimately are not dynamic; the end product is ostensibly as static as the fingerprints to which they are compared. Before we learn anything about the technical aspects of pod-replication, we are privy to several replications in process. In each example, replication is a dynamic process, a crafting, that culminates in a static product. "I've watched a man develop a photograph," Miles narrates upon finding an incomplete pod-person. "He dipped the sheet of blank sensitized paper into the solution, slowly swishing it back and forth, in the dim red light of the developing room. Then, underneath that colorless fluid, the image began to reveal itself—dimly and vaguely—yet unmistakably recognizable just the same" (59). The photograph analogy is complemented by several other such descriptions: a medallion lacking its final defining impression (36), a doll "made by a primitive South American people" (98), and an artist's "preliminary sketch" (61). In each comparison, the initial impression—the first rough cut—makes the object recognizable, defines the object as an image, a coin, a doll, a portrait. The secondary marks individualize the object. "There on that shelf lay Becky Driscoll—uncompleted," Miles explains when he finds Becky's doppelganger in the process of becoming human. "There lay a . . . preliminary sketch for what was to become a perfect and flawless portrait, everything begun, all sketched in, nothing entirely finished" (61). Once finalized, these things (a medallion, a sketch, a doll), and by extension the pod-people they metaphorically represent, become static objects: a coin that can only change its shape by being melted and destroyed, a portrait that captures but a second, an aspect of the subject's personality, a doll that represents humanity without being fully human. We might even say that Galton's description of a fingerprint as fixed and unchanging has been literalized as it is internalized: once formed, pod-people do not grow or change; indeed, they have a five-year lifespan—whether they replicate the bodies of growing children or elderly adults.

As the body's individuality becomes more and more atomized in the narrative, identity begins, deceptively, to appear quantifiable, tangible, and self-evident. Likewise—as we saw in chapter 3—aspects of personality, mind, emotion, and thought take on material form. Which brings me back to the novel's resident psychologist, Mannie Kaufman, who claims that the replication process includes "thought, memory, habit, and mannerism" as transferable biological substrates: "When you wake up, you'll feel just exactly the same. You'll be the same, in every thought, memory, habit, and mannerism, right down to the last little atom of your bodies. There's no difference. None. You *are* just the same" (171). That Mannie's description figures psychological substrates as atomic and material is not shocking in and of itself: psychologists, including William Marston whom we met in chapter 2 and the telepathic researchers we encountered in chapter 3, believed that thought could be represented, rendered, or understood as material. Marston even invented a term, the *psychon,* which rivaled the *neuron,* and was intended to represent psychology's smallest unit of investigation. What remains problematic is Mannie's assurance that because the replication is atomic, "there's no difference. None. You *are* just the same."

Although the pod-people, in the guise of human friends, neighbors, and associates, argue that there is no change, that the cellular pattern is expertly and completely replicated, the duplication is never as perfect as the alien spokespeople claim. What is missing is not a physical marker but an affective trace: the intensity that would normally accompany socialized emotions. Teresa Brennan uses the phrase "transmission of affect" to indicate the process of empathetic affective exchange between people and their environments, and the phrase is also quite literally useful to a discussion of the pod-people's replication practices. According to Brennan, "the transmission of affect was once common knowledge; the concept faded from the history of scientific explanation as the individual, especially the biologically determined individual, came to the fore" (2004, 2). As we have already seen, the rise of the biologically determined individual coincided with the development of various biometric identification systems, including fingerprinting. With the rise of the individual, the transmission of affect held less scientific weight. As if to literally enact this failure of transmission, affect is not transferable between the original human body and its pod replica in *Invasion of the Body Snatchers.* Indeed, a flattened affect is effectively the only way to detect a deceptive body—one that has been otherwise flawlessly replicated.

One way to parse out this flattening is through Eric Shouse's distinction between emotion and affect. According to Shouse, "an emotion is the projection/display of a feeling. Unlike feelings, the display of emotion can be either genuine or feigned," while "an affect is a non-conscious experience of intensity" (2005). Shouse explains, in accordance with Antonio Damasio, that "without affect, feelings do not 'feel' because they have no intensity" (2005). The pod-people display preexperienced thoughts and memories along with learned behaviors, habits, and mannerisms, all of which are repeatedly performed over time. What Miles terms emotion is better described as affect, particularly when he describes what the pods lack: "There was no *emotion*, not really, not strong and human, but only the memory and pretense of it, in the thing that looked, talked, and acted like Ira in every other way" (Finney 1996, 182). Replacing *emotion* with *affect* reveals that the pod-people are perfectly capable of feigning emotion, "the memory and pretense" of feeling, without the affect behind it: "there's no real joy, fear, hope, or excitement in you, not any more. You live in the same kind of grayness as the filthy stuff that formed you" (182).

Miles discovers this difference for himself on several occasions: after visiting Professor Budlong's house, he comments on the stacks and stacks of yellowed papers that were once part of a thriving research agenda; when Miles confronts the replicas with their lack of procreative ambition, their lack of love: "Can you make love, have children?" Mannie replies, "I think you know that we can't, Miles. Hell . . . you might as well know the truth; you're insisting on it. The duplication *isn't* perfect. And can't be" (Finney 1996, 183). Nor is Miles alone in his diagnosis; Becky, who has returned to town following her divorce, confides in Miles that her Uncle Ira seems different—strange. Although Ira walks, talks, looks like, and remembers everything her uncle should, Becky feels that he is different. "With this—*this* Uncle Ira, or whoever or whatever he is," Becky argues, "I have the feeling, the absolutely certain *knowledge*, Miles, that he's talking by rote. That the facts of Uncle Ira's memories are all in his mind in every last detail, ready to recall. But the emotions are not. There *is* no emotion—none—only the pretense of it. The words, the gestures, the tones of voice, everything else—but not the feeling" (21; emphasis added to *knowledge*). What Becky knows to be true is not something she can measure, but something she feels. As we have already seen, what differentiates the original Ira from his double is not his fingerprints or any other biometric data, but the lack of affect, the lack of feeling behind emotional expressions.

This marked lack of affect, even in the presence of feigned and/or habitual emotion, is best illustrated in Miles's clandestine encounter with a group of replicas who do not realize they are being observed. The aliens enact a parody of their own earlier conversations with Miles and

Becky by repeating scripts and catch phrases verbatim, but with a mocking edge. "Now the little old lady's voice deepened. 'Don't bother to explain, Wilma'—she was imitating my tone and manner to perfection . . . Then they all laughed—soundlessly, their lips pulled back from their teeth . . . and I knew these weren't Wilma, Uncle Ira, Aunt Aleda, or Becky's father, knew they were not human beings at all" (Finney 1996, 136). Miles account confirms that the pods are expert impersonators. But their act has little if any intensity or affect behind it. Indeed, their emotion is signified by a collection of physiological movements: "their lips pulled back from their teeth."

Within the limited scope of the narrative, affect does not fit into the code of the fingerprint because it is in excess of that classificatory scheme. It is not the marker but the missing marker, an identifiable absence. Through the detection of affect (and/or its lack), the protagonists create a novel identification method that challenges the fingerprint as marker and code for bodily knowledge production. The fingerprint remains a visceral signifier throughout the novel—a code whose absence is impossible to imagine—but it is the juxtaposition of fingerprinting and affect that throws the fingerprint's code into relief, creating a sense of estrangement from an infamous classificatory scheme. So much so that when we are told by Budlong that brain wave patterns have always existed, that "they weren't invented, only discovered. People have always had them, just as they've always had fingerprints" (175), we would do well to remember that systems of identification are constituted and employed for particular scientific and cultural purposes. Or, as in Simon Cole's larger narrative about the rise (and fall) of the fingerprint as marker, the novel illustrates that "identification methods do not flourish and become widely accepted solely on technical grounds. The acceptance of a new identifier as useful and reliable occurs within a particular social, cultural, and historical context" (Cole 2001, 293). For Miles, Becky, and the Belicecs, fingerprinting fails them at the moment of their greatest need. Doppelgangers overwhelm classification by becoming unremarkable; the self-signatures of the body fail to provide unique markers. The affective trace, while not theorized or substantiated by the scientists of the novel, becomes a new code for identification.

This is not to suggest that the affective trace is a better or more useful classificatory system than the fingerprint, that the fingerprint as code has vanished from our cultural milieu, or that the false binary of material/nonmaterial markers (created by the novel) is one we should maintain. Instead, I hope to have demonstrated that the code of the fingerprint, like Wegenstein's "faciality," has introduced and, in turn, benefited from, three assumptions: the body is self-reporting, individuation has been internalized, and the body remains a fulcrum for knowledge production *and its undoing.* Each of these assumptions is also crucial to an understanding of the technocultural history of lie detection, insofar as we assume that deception can be read from a self-reporting body, that deception has been internalized and written on multiple sites within the body, and that the body that produces knowledge of deception can also thwart its detection. As I explain in the next chapter, many of these cultural assumptions about fingerprinting as code continue to inform contemporary sciences of lie detection, including Brain Fingerprinting, an EEG technology purported to catalog the unique contents of one's memory.

Positioned as it is in the 1950s, at the liminal point between fingerprinting's colonial origins and post–9/11 resurgence, *Invasion of the Body Snatchers* reminds us that our cultural fascinations with marking the Other and with identifying deception have many histories, including the literary. And here, as in previous chapters, the body both produces and challenges knowledge production. In an age of terrorist threats, immigration disputes, and identity theft, we have retreated to systems of identification that rely on the visual, visible differences between "us" and "them"—techniques whose colonial origins inflect their current use.

While Congress did not pass the universal fingerprinting bill of 1940, or the Citizens Identification Act of 1943 (Cole 2001, 249), we are once again on the verge of biometric national identification cards in the United States. In fact, the Senate passed the Real ID Act in May 2005; it was initially slated to be implemented in May 2008. The new system's purpose is to impede illegal immigrants from obtaining work, boarding planes, and filing taxes; once implemented, the act would also impact U.S. citizens who will be required to present a birth certificate (along with other documents) when obtaining a driver's license. The Real ID Act begs questions about state surveillance and the ways in which biometrics are still being used as a means for identification. Abroad, Thailand has adopted biometric national identification cards, as have France, Germany, and Singapore. What has been lost in the panic of terrorism and war is the cultural cachet of fingerprinting—the historical development of a biometric technique and the construction of a culturally resonant code.

# A Tremor in the Brain

## fMRI Lie Detection, Brain Fingerprinting, and the Organ of Deceit in Post–9/11 America

Lies, damn lies—the prints are all over your cortex.
—LONE FRANK, *Mindfield: How Brain Science Is Changing Our World* (2007, 253)

In his 1996 novel *The Truth Machine,* James Halperin predicts that an altruistic child prodigy will design a foolproof and widely deployable lie detector by 2024.[1] The Armstrong Cerebral Image Processor (ACIP) will be based on a "combination of physiologically enhanced MRI and cerebral image reconstruction" (176). This "Truth Machine" would be a far cry, technologically speaking, from traditional polygraphs that detect deception by monitoring various physiological processes (including heart rate, blood pressure, and respiration) and equating changes in these processes to emotional stress and therefore deception. Indeed, Halperin's novel foretells a new class of lie detector: one that measures prevarication via brain imaging. Over the past decade—and particularly in response to the attacks of September 11— Halperin's vision has become a reality as scientists developed detectors based in the brain imaging techniques of functional magnetic resonance imaging (fMRI) and electroencephalography (EEG).

In the wake of these new detectors, the polygraph has been both demonized and ostensibly outmoded. The new technologies are said to offer all of the benefits of lie detection with fewer drawbacks. However, as I have been arguing, brain-based detectors not only are related to assumptions underlying historical instruments but also are products and perpetuators of their genealogical traditions, including the dynamic relationship between literature, science, and technology; the theoretical construction of the deceptive consciousness within laboratory science; the literacy of mind reading via the visualization of thought; and the code of the fingerprint. While this chapter begins to draw together the many threads of argumentation running through this book, you will find more satisfying knots tied in the coda which follows. In this final chapter, I introduce one last set of terms and concepts. Here, I contend that the discursive juxtaposition of polygraphy and brain-based detection is a representational strategy that foregrounds the corrective advantage of brain-based techniques, creates an artificial rupture between contiguous technologies, and ignores the shared assumptions foundational to fMRI, EEG, and two older "truth telling" technologies (polygraphy and fingerprinting), while also masking a complex genealogy of theoretical issues that I have been addressing throughout this book. The unfounded privileging of new technologies has been highlighted in STS scholarship, which "has tended to prioritize change and innovation rather than studying how dominant groups continually prevail at embedding meanings that maintain their continued power within technologies" (Campbell 2005, 395).[2] Through a case study of post–9/11 scientific experiments with fMRI and EEG, I highlight, historicize, and analyze the "embedded meanings" in the representation of brain-based detection.

After briefly explaining the 9/11 context for the emergence of brain-based detection, I address three pertinent but often overlooked assumptions that undergird fMRI and Brain Fingerprinting: first, brain-based detection simultaneously creates and describes the brain as the organ of deceit,[3] establishing quantifiable connections between brain and mind, biology and behavior. Proponents of brain-based detection attempt to rhetorically distance themselves from the body as it was examined in traditional polygraphy and fingerprinting in order to more accurately record the secret interiority and intentionality of individuals; the result is not a shift in ideology but a repackaged psychophysiological approach to a long-standing construct of the mind-brain, what I term the *biological mind*. Second, brain-based techniques ground cultural norms, such as conceptions of truth and morality, in biology.[4] By isolating deception and locating its sources in the brain, brain-based techniques reintroduce biology as the static building block for cultural conceptions and valuations of truth and morality. Alterations in the brain's hemodynamics and electrical activity become signifiers of deception's physiological origins. Finally, as truth is essentialized as bodily default, deception is interpreted not only as morally wrong, but as a sign of biological deviance.

## Technological Exigency and the Guilty Knowledge Paradigm

Exigency is typically defined as one's reason for speaking, for making arguments, for seeking solutions. Proponents of fMRI lie detection and Brain Fingerprinting often compare their techniques to polygraphy; however, they have found exigency for the further development and implementation of brain-based detectors based on the cultural context of post–9/11 America. In this section, I describe the brain-based technologies of fMRI and Brain Fingerprinting and explain the shifts in their scientific and journalistic presentation post–9/11.

Based on electroencephalography, Brain Fingerprinting was developed and patented by Lawrence Farwell, who, like William Marston, is a Harvard-trained psychologist; Farwell founded Brain Fingerprinting Laboratories in 2003. While Farwell is only one of several researchers working on the forensic application of EEG, he is the most prominent researcher in the national media due in large part to his successful self-promotion. His patented technique is worth examining given the persuasive and ubiquitous characteristics of his arguments concerning the technology.

Farwell's technique measures the electrical activity of the brain as an individual is exposed to a stimulus. To test a subject, Farwell fits him with a helmet apparatus that contains multiple electrodes and asks him to focus his or her attention on a screen where various pictures, words, and sounds appear. Farwell requires that each subject respond on a hand-held keypad whenever she sees a "target" appear on the screen. She may also be asked to click a button if s/he recognizes any of the other stimuli; however, her conscious responses are not necessary for the test to work. Indeed, Farwell's Brain Fingerprinting technique measures the P300 brain wave—an electrical signal that is ostensibly out of our conscious control. The P300 wave is said to spike 300 to 800 milliseconds after subjects "recognizes and processes an incoming stimulus that is significant or noteworthy" (Farwell 2001a). According to Farwell, Brain Fingerprinting produces not only a record of stimulus recognition but a catalog of the "information stored in the human brain" (2001a). Akin to its biometric namesake (fingerprinting), Brain Fingerprinting allegedly provides a means to mark and track individuals by producing a record of one's individuality—in this case, individuality as defined by the things that one recognizes and/or knows.

An outgrowth of magnetic resonance imaging (MRI) technology, blood-oxygen-level-dependent functional MRI (BOLD fMRI) uses electromagnetic waves to scan subjects' brains for changes in blood oxygenation levels. Scientists hypothesize that neural activity requires oxygen, which neurons cannot independently produce; active neurons receive more oxygen from the blood than inactive neurons. BOLD fMRI is sensitive to this variation in blood oxygenation levels; increased brain activity is inferred from increased blood oxygenation.[5]

To measure which, if any, regions of the brain are activated by deception, many fMRI studies ask participants to commit a mock crime (similar to those I described in chapter 2), and then lie about their activities while inside an fMRI machine. A computer asks questions, and subjects are asked to respond by pressing "yes" or "no" on a handheld keypad. Because the technique constantly monitors brain activity, researchers claim that deception can be detected even before the subject presses an answer into the keypad. Thus far, the anterior cingulate gyrus and the prefrontal and premotor cortex have been identified as "more active during Lie than Truth" (Langleben et al. 2002, 731).

In terms of protocol, fMRI detection and Brain Fingerprinting rely on similar test paradigms to contemporary polygraphy. In both fMRI and Brain Fingerprinting the mental activity often is being tested is memory: the recall of various stimuli in Brain Fingerprinting and the recall and subsequent inhibition of past events in fMRI deception experiments. Any conscious denial of a memory is deemed indicative of guilt or deception, particularly if either test ostensibly detects familiarity with a particular stimulus. The two newest detectors test a subject's memory via the established protocol of the Guilty Knowledge Test (GKT).[6] The GKT, which has been used in traditional polygraphy since 1959,[7] is made up of probe, target, and irrelevant stimuli (words, images, pictures, etc.); irrelevant stimuli establish a baseline from which to test reactions to target and probe stimuli. Probes are crime-relevant information, the possession of which implies "guilty knowledge" and therefore criminal involvement. Targets include publicly available crime-relevant information (from trials, the press, etc.) that allows technicians to establish a baseline response (autonomic or brain-based) for recognized but not self-incriminating or "guilty" knowledge. The GKT assumes that only subjects involved in the crime would recognize and unconsciously react to the probe stimuli. In other words, only "guilty knowledge" will evoke reactions in the region of the body (now brain) being monitored.

Within the rubric of the interrogation, "guilty knowledge"—like William Marston's "deceptive consciousness"—is produced in part by the very test designed to ferret it out. The GKT test works by contrasting reactions to "targets," "irrelevants," and "probes" the content of which has been determined before the test is administered. Given the importance of concepts like "guilty knowledge," detection of any kind is not only about lying after having committed a crime; it constitutes a form of ultrasurveillance that is invested in uncovering the concealment of experiences deemed to be deviant. As we shall see in later sections, this concept of "guilty knowledge" affects not only testing protocols but the language used to delineate contours of the biological mind.

Despite the fact that both polygraphy and brain-based detection rely on a similar protocol, proponents of Brain Fingerprinting and BOLD fMRI detection separate their techniques from traditional polygraphy by naming and lingering over the latter's ostensible flaws. Oft-cited defects include that polygraph machines measure and draw conclusions from peripheral (rather than central) physiological phenomena such as blood pressure, respiration, and heart rate; that the test cannot always identify lying suspects, particularly pathological liars and those trained to overcome the protocols; and that lie detector results are not regularly accepted in U.S. courts, and, more generally, cannot always be supported by physical evidence in the absence of DNA and fingerprints.[8] Researchers and reporters alike consistently describe and characterize brain-based detection as a corrective to these flaws. Daniel Langleben argues that "fMRI is a more direct marker of brain activity than the polygraph. The polygraph is based on the changes in the heart rate, blood pressure, breathing, and the electrical conductivity of the skin that often occur when people are lying. These measures reflect peripheral rather than central nervous system activity and may vary widely across individuals" (Society for Neuroscience 2002).[9]

Post–9/11 anxieties have made brain-based research even more relevant and desirable.[10] Major newspapers have characterized fMRI and Brain Fingerprinting as "a logical next step in security" (Streitfeld and Piller 2002). Researchers have also used the post–9/11 context to justify new lines of research, arguing that "the development of a technical means to detect deception that is superior to the polygraph is of critical importance in today's national security environment" (Happell 2005,

683). Similarly, Lawrence Farwell notes, "Before 9/11 it was clear we could detect whether somebody was an FBI agent or not. The response of many people in that field was, 'Ho-hum, that's nice.' After 9/11, it becomes extremely relevant whether somebody has been through an al-Qaeda training camp or whether he's a relief worker who has just been in the same area" (Gammage 2002). In its introduction, Daniel Langleben et al.'s 2005 scientific publication alludes to 9/11 as a rationale for brain-based research. He notes that "recent changes in the defense priorities of the industrialized nations have increased an already strong demand for an objective means of detecting concealed information" (262). Although Langleben et al. do not return to this concept in their conclusion, connections to the U.S. government, Homeland Security, and the military abound in brain-based detection research. Most of the initial research was funded by grants from the CIA, FBI, and Department of Defense.[11] This latest experiment received funds from the Defense Advanced Projects Agency through the Army Office of Research. Not only does 9/11 provide exigency, but it helps mask several foundational assumptions of brain-based detection that firmly link the technologies to the body.

## The Organ of Deceit

Despite representations to the contrary, brain-based detection is as invested as traditional polygraphy in correlating deception and truth with the body. The difference is the location of measurement: brain-based detection focuses on changes in the central instead of the autonomic nervous system. Specifically, researchers are casting their mechanical gaze—their Münsterbergian mental microscopes, as it were—on "the biophysical seat of decision making" (Happell 2005, 673) or "the organ that produces lies, the brain" (Ganis et al. 2003, 830).[12] The ostensible shift from periphery to center is crucial to the discourses used to describe brain-based lie detection, but it is also the illusory product and producer of assumptions about the brain's centrality to identity and its proximity to mind. When researchers argue that brains are the central hub of deception they construct the brain as both locus of mind and obliging organ, more compatible with scientific inquiry than the variable and suspicious body.

Let me begin with that "organ of deceit," which is more aptly characterized not as the brain, but "the mind-in-the-brain" (Beaulieu 2000), or what I have termed the *biological mind*. Thanks, in part, to the biological turn in the cognitive sciences, the mind has become an "informational object" (Beaulieu 2004, 371), one more system that could "be understood as a range of quantifiable events, a limited scale of possibilities" (Stam 1998, 2).[13] As a concept, the *biological mind* attempts to encapsulate this turn and its impacts on brain-based lie detection. As I define it here, the biological mind is a hybrid of physiology and psychology: it is both a mass of quantifiable data—a fleshy and obliging organ, compatible with techniques of scientific measurement—and also the last bit of matter between science and the mind. While the biological mind cannot merely be defined as or by its material components (cells, neurons, axons, dendrites), the biological mind assumes that the brain is so closely allied to the mind that its physiological structure and activity are indicative of its psychological states.[14]

That we read the brain for clues about everything from psychology to behavior is, at least in part, a consequence of the literacy of mind reading I discussed in chapter 3. The science fiction of the 1930s through 1950s envisioned mechanical access to the mind via technologies that were mixtures of EEG, telepathy, polygraphy, and imaginative extrapolation. Translated for the new detectors, this literacy results in reading practices that elide mind, brain, and behavior. For a more contemporary example, take Arthur Caplan's concerns about neuroscientific imaging technologies: "unlike the link between one's genes and one's behavior, which is 'fairly complicated,' he says, 'the link between brain and behavior is pretty tight, so it becomes more problematic when you start to have information in real time about the brain'" (quoted in Frith 2004). At first blush, this statement from the director of the University of Pennsylvania's Center for Bioethics and Trustee Professor of Bioethics in Molecular and Cellular Engineering seems suitably cautionary. And yet, the reason we need to exercise caution relies on an assumption about "the link between brain and behavior" that is "pretty tight."

As I illustrate in this section, discourses surrounding brain-based lie detection—and the biological mind—show that neuroscientific research has shifted the site of access to consciousness without reevaluating the body-brain-mind hierarchy. It is, as Elizabeth Wilson argues concerning neuropsychology, a "Cartesianism that has been repositioned but not resolved" (1998, 124). Imaging technologies reinforce the dualism of contemporary neuropsychology—a science that "rescues only the central nervous system (and then only a small part of that) from Cartesianism"

(123) by transferring psychophysiological measurements from the body to the brain (and central nervous system).

Brain-based detection often redefines the "physiology" of its psychophysiology by elevating the brain, by marking and marketing it as a more direct indicator of consciousness than secondary autonomic markers.[15] Nonetheless, like their polygraph counterparts, brain-based techniques are *indirect* measures of mental and emotional states based on changes in physiological processes. Functional magnetic resonance imaging (fMRI) tracks changes in neuronal activity, for example, by monitoring changes in the blood-oxygen-level-dependent (BOLD) signal.[16] Brain Fingerprinting (like its parent science EEG) measures changes in the brain's electrical activity.

Moreover, even as brain-based detection attempts to excise the body—to make it seem incidental—studies have illustrated the way the body can affect the collection and interpretation of data. On the most basic level, a subject's movement while in the fMRI machine can cause motion artifacts, the corruption of fMRI images. More important, normal and pathological aging may also affect neural activity, thereby necessitating different interpretive strategies for fMRI data, as reported in a 2003 study on BOLD fMRI.

> Caution must be taken to avoid misinterpreting the results of BOLD fMRI studies. The BOLD signal reflects the influence of neural activity on [cerebral blood flow], and therefore development-, age-, or disease-related changes . . . might influence our ability to attribute BOLD signal changes to alterations in neural activity. (D'Esposito, Deouell, and Gazzaley 2003, 871)

In short, bodies affect brains and impact data collection in BOLD fMRI. Indeed, recent neuroscientific research has also hypothesized that the body may very well have a certain say in how brain processes are carried out. Research on patients with brain lesions has led Antonio Damasio, professor of neuroscience at USC, to argue in his popular publications that "the body contributes more than life support and modulatory effects to the brain. It contributes a *content* that is part and parcel of the workings of the normal mind" (1994, 226). The body and brain are in constant communication; "the representations your brain constructs to describe a situation, and the movements formulated as a response to a situation, depend on mutual brain-body interactions" (228).

Despite this dynamism, brain-based lie detection research often relies

on the long-standing, implicit, and still influential assumptions of locationalism: "The search for localized regions tied to specific functions . . .
a continued search for a substratum of mind, using anatomical/physio-
logical techniques . . . [and] the division of labor between 'mind' lines of
work and 'brain' lines of work continues to be strongly represented"
(Star 1989, 182). Indeed, both polygraphy and brain-based detectors—
like many imaging technologies—sustain a mechanistic and locational
understanding of the body and its organs, including the brain. Polygraphy assumes a correspondence between physiological changes and emotion; similarly, brain-based techniques such as EEG and BOLD fMRI assume that there is a correlation between physiological changes and
decision making (particularly the inhibition demanded by deception).
As I argued in chapters 2, 3, and 4, connections between the body and
mind are far more dynamic than these techniques imply. Yet, the
metaphorical construction of the biological mind enables researchers to
exteriorize, codify, and classify interiorities, to extrapolate mental activity from the brain's mapped, and therefore predictable, physiology.

Spatial metaphors in particular, like "mapping" and "storage," make
the theoretical and intangible more concrete by eliding the difference
between mind and brain.[17] According to Beaulieu, "Maps link the life of
the mind and the space of the brain" (2003, 562), changing the basic co-
ordinates of research from theories of mind to atlases of biology. Documents written by Lawrence Farwell and made available on the homepage
for Brain Fingerprinting Laboratories, for example, liken the brain to a
filing cabinet or a "hard drive" capable of storing and retrieving information (Brain Fingerprinting Laboratories 2001b). Implied here are
several assumptions about brain function, information processing, and
memory, which Farwell addresses in an in-house document entitled
"Brain Fingerprinting Testing and Memory Issues." "Under normal circumstances," he argues, "one's memory for significant events (such as
committing a major crime) is intact, even long after the event" (2001b).
Farwell's arguments and his choice of "fingerprinting" as a metaphor
links Brain Fingerprinting to its biometric namesake as a means of char-
acterizing the nature, accessibility, and retrieval of information written
on—or in this case *in*—the body. As we saw in chapter 4, Francis Galton,
who helped bring fingerprinting from colonial India to Scotland Yard in
1901, argued that the ridged markings of our digits are unique indica-
tors of an individual's identity that neither fade nor change over time.
Similarly, Farwell's premise is that each brain contains unique informa-

tion that is attached to different memories; thus, the term *Brain Finger-printing* implies that each brain represents a unique map of life experiences, including any crimes or wrongdoings committed.

Other brain imaging technologies such as fMRI make more explicit use of mapping and mapping metaphors in their construction of the biological mind. Ideally, BOLD fMRI detection will not only isolate prevarication within individual brains but help scientists identify invariant cerebral markers of deception. The latter is a response to one of the failings of traditional polygraphy—one that even William Marston was attempting to eliminate in his laboratory experiments: the interindividual variability of the autonomic nervous system (1923). "Since deception-induced mood and somatic states may vary across individuals," argue Langleben et al., "a search for a marker of deception independent of anxiety or guilt is justified" (2002, 728). As we shall see in a moment, the desire to find deception in its purest form and, thus, the choice to separate reason from emotion (the act of deception from any anxiety or guilt) do not account for several theories in the neurosciences that integrally link reason and emotion (Damasio 1994).

Furthermore, spatial markers of deception are problematic for many reasons: first, deception is a working object; it varies between laboratories (whose researchers use different protocols for data collection and representation); individuals; social groups; and cultural and historical settings (Bok 1978). As we saw in chapter 2, Marston's introspective records reveal a wide range of individual definitions of and reactions to deception. Second, fMRI—like other brain imaging technologies—produces the very structures it describes by translating the brain into discrete and measurable units that do not naturally exist in the brain. The map, as Anne Beaulieu argues, "is an effective metaphor for the coordination of systems of knowledge on which atlases rely, while naturalising them as features of the territory of the brain" (2004, 378). Mapping implies a fixed terrain in which certain landmarks remain stationary, routes to and from locations do not often change, and activities associated with certain areas are well-established. Moreover, a map is a tool, often used to delineate a singular purpose, like a trip to the Grand Canyon. When charting such a journey, the rest of the country, and the world, tends to fall away. Mapping implies a certain sameness, a universal landscape: you might find more or less detail, a difference of scale, or markings for differing activities and audiences, but no matter where you purchase your atlas, the general layout of the land is consistent from one map to another.[18] Taken together, these critiques of spatial markers for deception reveal assumptions that are built into fMRI detection. Like the positron-emission tomography (PET) scans detailed by Joseph Dumit, fMRI could also be said to "build these neuroscience assumptions into its architecture and thus can *appear* to confirm them, while necessarily reinforcing them" (2004, 81). Representations of anatomy or physiological processes, then, do not merely reveal the biological mind, they help define its features and function.[19]

One such feature that is consistently highlighted in brain imaging literature is the claim that the brain is, apparently, under less conscious control than even the autonomic nervous system, whose fluctuations can vary between subjects. "It is likely," notes one group of researchers, "that a subject cannot mask functional MR imaging brain activation patterns" (Mohamed et al. 2006, 685). Acting as the medium of consciousness or as consciousness materialized, the brain represents some truth about the subject that cannot be manipulated, masked, or denied. This "'normal' brain-type is the one that is, so to speak, passive and lets the real self talk through it" (Dumit 2004, 163). We encountered similar assumptions about what I termed the "self-in-translation" in chapter 3 via the literacy of mind reading. In the science and science fiction of the 1930s through 1950s, thought is figured as a material substrate that, like the physical body, can be measured—even and especially against one's will. Likewise, we might draw connections between the self-signatures of the body (chapter 4) and what I would call the self-signatures of the biological mind. As with the fingerprint and the mechanical inscription of emotion, the self-signatures of the biological mind appear not only to write themselves but also to be free from conscious manipulation. Langleben and Farwell argue that the brain's electrical activity and hemodynamics cannot be purposely manipulated by participants or suspects. Subjects may hold their breath, step on a tack, bite their tongue, or attempt to slow their heart rate, but none of these tactics that potentially thwart the polygraph will have any effect on brain-based techniques. In short, "faking cerebral activity to avoid the detection of deception is not feasible" (Lee et al. 2002, 157), or as Farwell puts it, "testimony may not be truthful, whereas the brain never lies" (Witchalls 2004). In each case, the relationship between mind, brain, and body is not cooperative, but antagonistic.

The procedure of pitting part or all of the body against the mind is also not new; in fact we saw several versions of this battle in Marston's conception of the deceptive consciousness (chapter 2) and the thought translations I analyzed in chapter 3. As a reminder, consider the following description of polygraphy from Allan Hanson's *Testing Testing: Social Consequences of the Examined Life* (1993).

> The two machines commune together as the polygraph reaches out to embrace the subject's body with bands, tubes and clips. The body responds loverlike to the touch, whispering secrets to the polygraph in tiny squeezes, twinges, thrills, and nudges. But both machines are treacherous. The body, seduced by the polygraph's embrace, thoughtlessly prattles the confidences it shares with the subject's mind. The polygraph, a false and uncaring confidant, publishes the secrets it has learned on a chart for all to read. The subject as mind, powerless to chaperone the affair, watches helplessly as the carnal entwining of machines produces its undoing. (93)

Hanson's description captures several crucial premises of traditional polygraphy—assumptions reinforced by the Cartesian dualism in which the mind is separable from the lowly, noncognitive body.[20] First, the body is "thoughtless" and mechanistic. Although it engages in a "carnal entwining" with the polygraph instrument, the body responds predictably, not as a mass of undifferentiated and fluctuating flesh. The mind, an immaterial "chaperone," remains permanently and impotently separated from the treacherous body, even as that body retains access to its secret confidences.

Brain-based detection may attempt to excise the autonomic body, but it does so by focusing on the brain, a bodily organ. Like the body of traditional polygraphy, the brain appears to betray the mind. In the case of BOLD fMRI detection, one's anterior cingulate gyrus is either "active" or "inactive"; in a Brain Fingerprinting test, the P300 wave shows whether information is either "present" or "absent." Explanations and testimony are neither necessary nor accurate once the brain—acting simultaneously as object and subject—has divulged its truth. Instead of a "tremor in the blood" (Lykken 1998), which is what polygraphs can be said to have measured, researchers are looking for a tremor in the brain. In the case of brain-based detection, though, this betrayal seems even more treacherous because the biological mind simultaneously plays two roles: that of immaterial subjectivity (mind) and unconscious organ (body/brain). As Paul Root Wolpe, senior fellow at Penn's Center for Bioethics, notes, brain-based technologies have a very specific impact on how we view ourselves "because in our [Western] culture, at least, we think of ourselves as brains. We think of our brains as the locus of our identity" (Frith 2004). Wolpe's remark echoes some of the assumptions we encountered in chapters 3 and 4 that the brain (and its brain waves) are the best and most accurate locus of identity. The brain may not be as "thoughtless" as the body, but it "prattles the confidences" all the same.

## Truth: The Brain's Default?

The turn to biology that constructs the brain as the accessible seat of deception also essentializes and biologizes truth and morality. In the experimental literature of brain-based detection, truth has been defined in terms of biology. Recent papers claim that "truthful responding may comprise a relative 'baseline' in human cognition and communication" (Spence et al. 2004, 1755), while "deception can be defined simply as 'denying what is'" (Langleben et al. 2005, 263). Discursively, these definitions of truth are products and producers of what have been called the efficient routes or hard-wired default of our brains. As a press release for the annual Society for Neuroscience conference, where Langleben first presented the results from his research in 2001, reports, a "new human brain imaging study shows that there may be an objective difference between lying and telling the truth that can be measured in the brain" (2002). In this section, I focus on the essentialist elements of fMRI research that posit truth and morality as the brain's default.

The process of uncovering nature's secrets is nothing new; neither are the consequences for debates about nature/culture. "In Western culture," argues Evelyn Fox Keller, "the threat or the allure presented by Nature's secrets has met with a definitive response . . . the scientific method—a *method* for 'undoing' nature's secrets: for the rendering of what was previously invisible, visible" (1992, 41). As we have seen thus far, brain-based detection is the newest method for rendering the organ of deceit (the biological mind) visible. But exactly what has been made visible remains to be seen. In the meantime, as brains are both natural objects and cultural/scientific constructs (Dumit 2004; Beaulieu 2002, 2003, 2004), we cannot avoid the matter of metaphors, which "powerfully redefines concepts like behaviour, nurture, culture, and environment" (Beaulieu 2003, 563).

When asked to describe their research in more general terms for the press, researchers' representations—and their uptake in the media—become more clearly constructive rather than descriptive. In an interview just after the attacks of September 11, Langleben tells a reporter that "if truth was the brain's normal 'default' response, then lying would require increased brain activity in the regions involved in inhibition and control" (O'Brien 2001). Less than a month later, when this story was picked up by an external news source, Langleben's statement is recast. He is quoted as saying that "when you tell a deliberate lie, you have to be holding in mind the truth" (Hall 2001); however, instead of letting Langleben's explanations stand, the reporter ostensibly conflates both quotes to argue that "being truthful is the brain's 'default' mode—not for morality's sake, but because it's less taxing on brain cells to come clean than it is to spin a convincing yarn while also hiding the truth" (Hall 2001). Fiction—the very ability to "spin a convincing yarn"—becomes an expendable and physiologically demanding phenomenon, while truth (and all of its associations: science, simplicity, objectivity, reality) is embraced as "less taxing." This logic of effort is curious, particularly because it is often paradoxically presented. William Marston made similar arguments, including the fact that "no man can lie without effort" (1938b); yet, as we saw in chapter 2, Marston's subjects often found lying to be more relaxing than truth-telling. Moreover, and as we saw in chapter 3, confessing the truth can require plenty of effort and be, arguably, as nerve-wracking as deception. Beyond this paradox, we need to delve into three seemingly unrelated issues if we are to further unpack Langleben's quoted statements as well as Carl Hall's reiteration, including theories of response inhibition, St. Augustine's conception of truth, and the dilemma of essentialism.

Thus far, fMRI detection research has analyzed experimental results using a response inhibition paradigm. Very basically, the two areas activated during deception (the anterior cingulate gyrus and prefrontal cortex) are active whenever the brain is involved in decision making—particularly when the decision involves the inhibition of a less desirable response (Frith 2004). According to Langleben et al., deception requires the inhibition of a default, though potentially damning response: the truth (2002, 731). "This concept suggests that inhibition of truthful response is a prerequisite of intentional deception" (727).

As with the mind/brain/body debate, theories of inhibition date back at least to the nineteenth century. As Roger Smith (1992) has cogently argued, nineteenth century conceptions of "inhibition" embodied cultural ideologies, including the self-regulation of conduct, humans' relationship to the natural world and instinctual behavior, and the relationship of an individual to his or her society. These same ideologies informed scientific theories about mind/body interaction, hierarchies of mind, and the regulation of various physiological forces and mechanisms. Indeed, inhibition has been imbricated in several paradoxes of philosophical and scientific thought: "The word evoked psychological explanation at the same time as it promoted the search for a physiological mechanism; it signaled the power of the mind over the body, yet it was part of the means by which the nervous system regulated itself; it implied purpose and value derived from mind, while it also described an order inherent in the fabric of nature" (Smith 1992, 16). In short, concepts of "inhibition" continually re-present "a dialogue between the desire to exercise moral control and the description of natural control" (10).

Langleben et al.'s claim concerning the inhibition of truth participates not only in the long-standing search for mechanical lie detection but also the social and scientific history of inhibition. Specifically, the elision of moral, psychological, and physiological regulation foregrounds—and complicates—the relationship between science and humanism. Although Hall's interpretation of Langleben's press interviews derides certain humanistic practices (including the creation of narrative fiction), Langleben et al. claim inspiration from a humanist: St. Augustine. Their interpretations are somewhat simplified, perhaps even scientifically streamlined, yet Langleben et al.'s reading of St. Augustine's philosophy is that "deception of another individual is the intentional negation of subjective truth" (2002, 727). In their 2005 article, Langleben et al. specifically refer to the following passage[21] from St. Augustine's exegetical treatise *De mendacio* (*On Lying*, 395 AD).

> Wherefore, that man lies, who has one thing in his mind and utters another in words, or by signs of whatever kind. Whence also the heart of him who lies is said to be double; that is, there is a double thought: the one, of that thing which he either knows or thinks to be true and does not produce; the other, of that thing which he produces instead thereof, knowing or thinking it to be false. (Augustine 1847, 384)

St. Augustine goes on to argue that "so must also truth be preferred to the mind itself, so that the mind should desire it not only more than the body, but even more than its own self" (395). Congruent with the implied message of Langleben et al.'s comment, St. Augustine's argument is not only more complex but morally charged. Throughout *De mendacio* and his other works, St. Augustine is invested in explicating man's relationship with God through scriptural exegesis.

Langleben et al.'s reference to St. Augustine jibes with other references to morality in contemporary scientific literature. In a survey of the early literature on brain-based detection by Spence et al., the authors conclude that "when humans lie they are probably using some of the 'highest' centres of their brains, a proposition that has implications for notions of moral responsibility" (2004, 1755). Spence et al. draw this conclusion based on the assumption that telling the truth requires less work from the brain: "In the normal situation the liar is called upon to do at least two things simultaneously. He must construct a new item of information (the lie) while also withholding a factual item (the truth), assuming that he knows and understands what constitutes the 'correct' information" (1757). The final clause of this statement reveals several key assumptions: truth is essential, definable, and universal, but only so long as subjects know, understand, and associate truth with the "correctness" of their response. As with William Marston's experiments, judging not only truth and, here, morality (our ability to judge right from wrong, or put another way correct from incorrect) depends on experimental protocols that correlate truthfulness with a known quantity.

Functional magnetic resonance imaging researchers' assumptions about truth's biological basis essentialize and universalize the nature and value of veracity; in so doing, they reveal the constructed character of the brain and behavior being studied. When we look outside of neuroscience to related disciplines, such as evolutionary psychology, we find sharply contrasting understandings of deception and the brain. In evolutionary psychology, for example, the brain is characterized as a piece of hardware whose primary function is to solve adaptation dilemmas for the sake of survival. According to this rubric, deception may very well be a positive, adaptive trait, not simply a physiologically wasteful activity.

In addition to being constructed by disciplinary ideologies, assumptions about truth's biological basis are circular. Akin to concepts like inhibition and locationalism, representations of the deceptive brain are presented as descriptors of natural phenomena, when, in fact, they are simultaneously constructs that reinforce the cultural and scientific ideologies that created them. This process distorts conceptions of both culture and nature: cultural constructions are elided with what is said to be natural, while "nature is being put to work in new ways that signal not so much its disappearance as its transmogrification" (Franklin, Luria, and

Stacey 2000, 19). One consequence of Langleben et al.'s logic, for example, is that cultural and moral conceptions of truth gain credence because they appear to be naturalized, essential components of our biology: our brains seem to not only have the "neural circuitry for deception" (Langleben et al. 2002, 727), but also a preference for efficiency, and therefore truth.

## Deception, Deviance, and Guilty Knowledge

The biological mind is represented as both readily accessible and essentially truthful; its measurable activities serve as the basis for hypotheses about behavior—including its evaluation and management. This brings me to my third and final point: in post–9/11 America, scientific and journalistic representations of Brain Fingerprinting have characterized guilt and deception as biologically measurable phenomena that pose a threat to the nation. Implicit in the presentation of brain-based detection is the hypothesis that if we defy our biological imperative for truth— if we lie and deviate from the experimentally defined norm of truth—science can objectively measure our deviation, perhaps even our propensity to commit future deceptive acts.

Instead of proving the novelty of their experimental techniques, researchers making such claims align themselves with both traditional polygraphy and centuries-old concepts of biological deviance. Polygraphy has always looked to the body for information about deception and deviance (Rhodes 2000; Alder 2007; Bunn 1997). The concept of biological deviance itself is not new; indeed, theories of biological deviance have played a large role in many criminological theories, including those of Cesare Lombroso, Francis Galton, and Havelock Ellis who each linked biological characteristics to (im)morality and propensities for deviance.

Despite these implicit connections to older technologies and theories, experimental and corporate literature for Brain Fingerprinting is invested in notions of biological deviance in two respects that I will unpack in a moment: (racial) profiling often determines the selection of suspect populations, thereby always already marking the deviance of certain bodies. In addition, the elision of data gathering and interpretation constructs Brain Fingerprinting as a technology that can derive intentionality from biological functioning.

Brain Fingerprinting relies on a profiling system that is reminiscent of its colonially developed namesake, fingerprinting. As we saw in chapter 4, fingerprints were first systematically employed by the British in Bengal, India, and thus began their existence as a progressive instrument for political control (Cole 2001; Sengoopta 2003). Through its classificatory system—and other established cultural systems of racism, including those that informed Luther Trant's lie detection practices in chapter 1—foreignness was often associated with criminality.[22] Fingerprints were used to track colonial subjects and criminal suspects, providing the British government and its citizens with one more physiological marker of foreign/criminal difference. Similarly, Brain Fingerprinting literature evokes the conflation between foreignness and criminality so evident in early biometric treatises, including various assumptions about common ethnic and national characteristics of terrorists—enemies who are almost always foreigners who manage to infiltrate the United States.

In one particularly salient example, Farwell argues that his technique can distinguish between the trained terrorist and the innocent international student: "A trained terrorist posing as an innocent Afghani student will have information regarding terrorist training, procedures, contacts, operations, and plans stored in his brain. Brain Fingerprinting can detect the presence or absence of this information, and thus distinguish the terrorist from the innocent person" (Cavuoto 2003). In a separate interview, Farwell makes a similar connection, arguing that the "difference between a terrorist who has been through a training camp and an Afghan student is the memory" (Spun 2002).[23] Note, of course, that neither example makes mention of domestic terrorists (like Timothy McVeigh), nor does it account for foreign nationals that come to America for other reasons, such as the pursuit of an education; instead Farwell's statement assumes that terrorists will only be found among foreign nationals or among terrorist cells in foreign countries.

Ironically, this brain imaging technology that is purportedly more objective than the polygraph continues to rely on and perpetuate a biologically collusive rhetoric that mimics the traditional assumptions behind lie detection: in this case, that guilt can be read from the body. Brain Fingerprinting Laboratory's corporate literature denies—even as it depends upon—various extrapolations of guilt or innocence. Several documents, for example, state that it is up to "a judge or jury to reach legal decisions such as whether a person is innocent or guilty of a crime" (Brain Fingerprinting Laboratories 2001e). Yet, in an interview for *60 Minutes,* Lawrence Farwell claims that "the fundamental difference between an innocent person and a guilty person is that a guilty person has committed the crime, so the record is stored in his brain. Now we have a way to measure that scientifically" (Brain Fingerprinting Laboratories 2001d). To justify these claims, Brain Fingerprinting lays claim to two foundational, though often unnamed and unjustified, assumptions: that there is a recognizable and testable difference between legitimate and illegitimate knowledge acquisition; and that the presence or absence of knowledge is indicative of intentionality and/or the actualization of this knowledge.

Procedurally, Brain Fingerprinting tests are touted as an objective measure for the presence or absence of information via the P300 wave; but how one acquires the knowledge is a far more important—and far more subjectively gathered—aspect of Brain Fingerprinting. Farwell claims, for example, that "by testing for specific information, Brain Fingerprinting technology can accurately distinguish between a trained terrorist and an innocent person who may have knowledge of certain locations, people and events for *legitimate* reasons" (Brain Fingerprinting Laboratories 2001c; emphasis added). In an extended passage (worth quoting at length for its use of repetition), Farwell explains how the larger testing procedure works to distinguish between legitimate and illegitimate—or deviant—knowledges.

> Any known *legitimate* means through which a subject may have encountered crime or terrorist-relevant information are examined prior to conducting a Brain Fingerprinting test. Standard protocols ensure that the individual has a chance to reveal any circumstances through which he may have had *legitimate* access to the crime-relevant information in question. Any information the suspect has obtained through *legitimate* means is eliminated from consideration before the test is administered. A suspect is tested only on information that he has no *legitimate* means of knowing, information he denies knowledge of, and for which he has no *legitimate* explanation if it turns out that the information is indeed stored in his brain. (Brain Fingerprinting Laboratories 2001c; emphasis added)

Here, in a subjective preliminary interview, the examiner is expected to sort, classify, and eliminate from the exam certain types of knowledge-acquisition situations. Farwell justifies the procedure not by explaining his ideological assumptions but by arguing for the specificity of Brain Fingerprinting's technique. He notes that "Brain Fingerprinting testing cannot be used for general, non-specific testing, something for which no reliable facts exist against which to compare the subject's answer" (Brain

Fingerprinting Laboratories 2001c). In short, a full investigation and subjective examination must be carried out *before* a test is ever ordered. To say nothing of scientific process, Brain Fingerprinting's reliance on a legitimate/illegitimate knowledge paradigm defies *due process,* by assuming a verdict of guilt or innocence long before the mechanical exam is ever undertaken.

Moreover, as carried out, Brain Fingerprinting extrapolates from knowledge acquisition to actualization—particularly in terms of post–9/11 counterterrorism initiatives. Instead of maintaining his claim that Brain Fingerprinting "does not directly detect guilt, innocence, lying or truth telling" (Brain Fingerprinting Laboratories 2001f), Farwell contradicts himself by arguing that a "Brain Fingerprinting test can determine with an extremely high degree of accuracy those who are involved with terrorist activity and those who are not" (Brain Fingerprinting Laboratories 2001c). On his "Counterterrorism Applications" Web page, Farwell claims to detect far more than "information present" or "information absent." Possible applications include:

- Aid in determining who has participated in terrorist acts, directly or indirectly

- Aid in identifying trained terrorists with the potential to commit future terrorist acts, even if they are in a "sleeper" cell and have not been active for years

- Help to identify people who have knowledge or training in banking, finance or communications and who are associated with terrorist teams and acts

- Help to determine if an individual is in a leadership role within a terrorist organization (Brain Fingerprinting Laboratories 2001c)

In this particular list, Farwell's focus shifts. Brain Fingerprinting tests no longer merely detect knowledge; they assess activity—even potential actions—through detection of "guilty knowledge." Farwell's technique not only describes knowledges, intentions, and activities but helps to produce and name them by redefining terrorism as perpetually latent, physiologically measurable deviance—a deviance that is accessible and measurable. It is made to sound like a physical characteristic akin to the centuries-old fingerprinting system of surveillance to which its name refers. The ability to distinguish between knowledges becomes a crucial element in the war on terrorism, not because knowledge is a new hazard in and of itself but because it aligns itself with a long tradition of science and biological deviance.

The emergence of brain-based detection has been hailed as the answer to polygraphy's perpetual problems. These new techniques ostensibly excise the autonomic body to focus on the central nervous system, require less subjective interpretation, and enable the visualization of deception's biological basis. As potential counterterrorism technologies, fMRI detection and Brain Fingerprinting have responded to the heightened anxiety in post–9/11 America. I have argued that the anxiety created by 9/11 has enabled a particular kind of presentation that privileges technoscientific advancement while returning to many of the assumptions foundational to traditional polygraphy and fingerprinting. In addition, given the biological turn of the cognitive sciences, brain-based detection proponents use arguments about nature to buttress and justify cultural norms. Far from describing the brain and its functions, the protocols of and assumptions about fMRI and Brain Fingerprinting produce and are the products of brain models that reintroduce and reinforce connections between biology, deviance, and deception.

Indeed, even the novel with which I began this chapter amplifies rather than simplifies the ambiguities of so-called scientific progress. While author James Halperin has invested in Lawrence Farwell's company, *The Truth Machine* does not unequivocally endorse the implementation of brain-based detection. In the novel's final scenes, the truth machine has clearly preserved the human race from self-destruction, yet, it has also effectively weakened it. While the ACIP provides security, it does so through an absurd—and somewhat dangerous—reduction of *chance*.

> In some ways, the Truth Machine's been a crutch. . . . I regard our *dependence* on it as its biggest drawback. Before the ACIP, we dealt with an incredible amount of uncertainty—both in our careers and personal lives. It takes a special kind of intelligence to deal with the possibility that every statement anyone makes to you might be a lie. Today things are a lot more cut-and-dried. The part of our brains that used to deal with uncertainty might've atrophied a bit, don't you think? Take away the Truth Machine and you're gonna see people who can't deal with life at all. (1996, 314)

We could read this passage in several relevant ways: first, the management of one risk inevitably engenders other hazards. In the novel, deception is constructed as a risk to public safety; lies are the root of crime, murder, and unhappiness. Because it detects lies and makes them public knowledge, the Truth Machine corrects the problem of deception. The irony (at least in the narrative passage quoted above) is that this risk management creates new problems: the atrophy of "a special kind of intelligence" that helps humans process the full range of possible human communication—including deception. Second, the simplification of one variable—like truth telling or the brain's biology—does not necessarily enhance our understanding and negotiation of human behavior. Instead, the result is a flattening of human interactions. Finally, *The Truth Machine* returns us to the biological mind with which I began, by relying on the locationalist assumption that there is a particular part of the human brain that is activated when we attempt to interpret differences between truth and lies: "The part of our brains that used to deal with uncertainty might've atrophied a bit."

Yet, despite the problematic claims of authors, scientists, and the popular press, brain imaging technologies are not inherently good or bad, advantageous or problematic. BOLD fMRI and EEG provide us with innovative tools to measure new aspects of the body in vivo. The challenge is that data produced by brain imaging technologies can be and have been further interpreted to explain behavior, personality, and identity (Dumit 2004). Thus, arguments about advantages and drawbacks cannot account for the complex interactions of culture, history, researchers, subjects, and technologies. As we have seen, fMRI detection and Brain Fingerprinting are enmeshed in arguments spun in a climate of nationalistic fear and anxiety that allows foundational assumptions behind polygraphy and fingerprinting to recirculate as uninterrogated ideology. The cultural context of the popularization of these brain-based technologies leads to a litany of repurposed, yet dynamic questions being posed by historians, legal scholars, theorists and neuroethicists—some of which have even been taken up in dialogue with the scientists themselves (Wolpe, Foster, and Langleben 2005; Conan 2006).

An analysis of scientific and journalistic presentations is another, crucial point of intervention that extends and complements calls for scholarship in STS literature. Brain-based detection, like other "suspect technologies" detailed by Campell (2005), offers "entry points for STS to contribute to struggles for social justice by producing useable knowledge about the scientific claims upon which technology is based, by situating them within the social context of their use and abuse, and by expanding the discursive space in which political interpretations are offered" (394). By bringing the discursive space of scientific publications—and their translation to popular media—under scrutiny, an analysis of language and context can strip brain-based detection of its appeals to innovation and technoscientific advancement, allowing us to examine its foundational assumptions.

While brain-based detection may allow us to manage one risk (the failures of polygraphy), BOLD fMRI and EEG have not resolved the underlying tensions between mind and body, intentionality and behavior, innocent and guilty knowledge. As Ulrich Beck notes, "science is *one of the causes, the medium of definition and the source of solutions* to risks, and by virtue of that very fact it opens new markets of scientization for itself" (1992, 155). Functional magnetic resonance imaging and Brain Fingerprinting have become marketable technologies not despite but because of their relation to the body found in polygraphy and fingerprinting. The simplification of variables, including the excision of the autonomic body, does not necessarily enhance our understanding and negotiation of human behavior; science may not be able to, and should not have to, provide any ultimate answers; our brains may not have a default.

# *Coda*

## Lie Detection as Patterned Repetition—From *The Demolished Man* to *The Truth Machine*

Meaning is not guaranteed by a coherent origin; rather, it is made possible (but not inevitable) by the blind force of evolution finding workable solutions within given parameters.

—N. KATHERINE HAYLES, *How We Became Posthuman* (1999, 285)

Truth be told, to observe the future without altering it is a scientific impossibility. But if your views remain fluid, even a false vision is far more valuable than no vision at all.

—JAMES HALPERIN, *The Truth Machine* (1996)

A book on the cultural history of lie detection could have begun with any number of milestones: Lombroso's work with the plethysmograph in the late nineteenth century, the establishment of the Scientific Crime Detection Laboratory at Northwestern in 1929, Wonder Woman's truth lasso (circa 1942), Fred Inbau's first disciplinary manual on the polygraph (1942), or even with the simple proposition that on or around 1917 *the* lie detector was invented. Instead of highlighting any particular historic event, I chose a genealogical approach that maintains my skepticism about establishing an official origin for lie detection technologies, while also illustrating the continued viability of lie detection in the American cultural imagination. My goal has been to contextualize the emergence of fMRI lie detection and Brain Fingerprinting, to demonstrate that lie detection technologies are infinitely adaptable, not despite but because new technologies rest on three foundational premises: that lies betray themselves through changes in our physiology, that bodies are self-reporting, and that lies demand the suppression of truth. In tracing these assumptions I have looked to the mutual imbrication of applied psychology, the law, and fictional marketing (chapter 1), the quantitative and qualitative inquiries into the deceptive consciousness via mock crimes (chapter 2); the literacy of mechanical mind reading (chapter 3); the history of fingerprinting as code for individuation (chapter 4); and the representations of brain-based lie detection in scientific and popular publications (chapter 5).

My hope, in completing this trek through decades and disciplines, is that the patterns of emergence and dissemination for various technologies of lie detection have become recognizable repetitions. As new lie detection technologies emerge, we would do well to not believe the hyperbolic claims that separate technologies from histories, from ideologies, and from previous incarnations. Contemporary lie detection is not standing on the shoulders of giants, nor is it a progressive alternative to some outdated, flawed technology. Instead, I have argued that, contrary to scientific and popular rhetoric, brain-based lie detection technologies are not immune to nor have they resolved the practical and ideological conundrums that informed lie detection at the turn of the twentieth century.

In this coda, I would like to reflect on the patterned repetition I discussed in the introduction through a brief comparison of two pieces of literature that played a role in *The Lying Brain:* Alfred Bester's *The Demolished Man* (1951/1953/1996) and James Halperin's *The Truth Machine* (1996). The first text concerns the exploits of a telepathic police force capable of "mind reading"; the second narrative envisions a near-future world (2024) in which the ACIP brain imaging technology has made lying virtually impossible. At first blush, these novels seem nothing like each other, particularly because one relies on the "pseudoscience" of telepathy, and the other depends on an accepted "science" akin to fMRI. Yet, as the preceding chapters have demonstrated, the different designations for scientific knowledge are somewhat arbitrary; they depend more on public and academic acceptance than some fact of nature or real accomplishment of technology. Moreover, the ways in which pseudoscience and the science figure access to the brain foreclose the ambiguity retained by fictional representations of telepathy and brain imaging.

Indeed, Bester's *Demolished Man* and Halperin's *Truth Machine* demonstrate in both form and content that peepers and ACIP scans are more alike than they are different. Thus, instead of remarking on the ways in which literature misrepresents, incorrectly predicts, or outwardly chastises (pseudo)science—positions that would unnecessarily limit the scope of this analysis—I have focused on the ways in which technology bridges and binds literature and science as marketers, interpreters, and arbiters of various instruments, techniques, and practices.

Despite the forty-five years between *The Demolished Man* and *The Truth Machine,* their plots are nearly identical: a CEO murders his competitor and avoids detection by reciting verse to foil the technologies of lie detection. While Ben Reich of *The Demolished Man* chooses an infectious advertising jingle, "Tenser, Said the Tensor," the main character in *The Truth Machine,* Pete Armstrong, relies on a piece of Americana: Walt Whitman's "Oh Captain, My Captain." What differs is not the method but the substance of their mental reiterations. The poems permit each man to avoid the national surveillance systems that are policing intent and deceit; however, one man hides behind utter banality, while the other screens himself in distinguished prose about an eminent man.

The different verses are, arguably, indicative of several aspects of lie detection we have seen throughout the past chapters: their ubiquitous repetition, which can be readily seen in any of the three assumptions I have been tracing; their progressive, even philanthropic appeal, which can be seen in the historical promise of the polygraph to mitigate the violence of the third degree and in contemporary lie detection's pledge to allay concerns about the accuracy and reliability of the polygraph; and their virulent hold on the American cultural imagination, which is evident in their revitalization in the face of technological failure.

Let us begin with repetition. In *The Demolished Man,* banality is a state of mind and the state of the universe: the continuous repetition of man's inventions and crises. "In the endless universe," explains the narrator, "there has been nothing new, nothing different. . . . This strange second in a life, that unusual event, those remarkable coincidences of environment, opportunity, and encounter . . . all of them have been reproduced over and over" (243). Ben Reich's struggle for capital and personal control, his psychological self-destruction, and his desire to thwart detection are nothing new. Neither is the tune he chooses. When he asks for the most persistent jingle from Psych-Songs, Inc., "a tune of utter monotony filled the room with agonizing, unforgettable banality. It was the quintessence of every melodic cliché Reich had ever heard. No matter what melody you tried to remember, it invariably led down the path of familiarity to 'Tenser, Said the Tensor'" (43).

> Eight, sir; seven sir;
> Six, sir; five, sir;
> Four, sir; three, sir;
> Two, sir; one!
> Tenser, said the Tensor.
> Tenser, said the Tensor.
> Tension, apprehension,
> And dissension have begun.      (45)

The familiarity of Reich's song—its contagion—is a result of form and content. Its rhyme and rhythm repeat melodic clichés with such tenacity that all songs collapse into its schema. The song's content, specifically the use of *tensor*, is inflected with universal mathematical algorithms, and not simply because the jingle was the theme song for a "flop show about the crazy mathematician" (42). According to the *Oxford English Dictionary*, a *tensor* is "an abstract entity represented by an array of components that are functions of co-ordinates such that, under a transformation of co-ordinates, the new components are related to the transformation and to the original components in a definite way." In short, a tensor is a vector that is consistent across different coordinate systems. In principle, it reminds us of scientific arguments for the persistence of certain physical laws across time, space, and history. Like the melody of the song itself—or a technology of lie detection—a tensor appears ubiquitous and predictable: an entity that reappears in similar form across multiple configurations.

In this respect, Reich's song illustrates the moments in which systemically employed lie detection (like a nation policed by telepaths) engenders further risks. As Ulrich Beck has illustrated, risk societies continually produce more hazards through the management of risks. Ben Reich's murderous impulses—his desire to thwart the system of Espers—is, at least in part, the result of the surveillance itself. If a tensor is a transcendent physical law, its ubiquity and its limitations combine to produce a certain tension within Reich's society. Perhaps "tension, apprehension / And dissension have begun" not in spite of but because of the peepers' systemic surveillance.

If Reich's verse illustrates lie detection's continual reemergence and limitations, Pete Armstrong's poem indicates their often progressive (and sometimes philanthropic) purpose. The main difference between the two men—and the two story lines—is the motivating force behind their desire to deceive. While Reich must thwart the Espers for selfish reasons (the maintenance of his company, for example), Pete fools his own truth machine "for the sake of the human race" (Halperin 1996, 340). Like those of his chosen poem's namesake, his actions are for the good of the nation. "Lincoln had long been Pete's historical inspiration," explains the narrator, "and Pete secretly hoped that if he succeeded in bringing the world a Truth Machine, future generations would venerate him as modern Americans revered Abraham Lincoln" (196). There is, of course, a slight megalomania here, similar to that seen in Victor Frankenstein, that allows Pete to reason that plagiarism, murder, and deceit are worth the potential benefits of a functional truth machine that could be implemented throughout society.

In a practical sense, he is quite right. The murder rate drops dramatically after the implementation of the device: "The national homicide toll in the year 2000 had exceeded 45,000" (Halperin 1996, 49), but "there were fewer than 500 murders in the United States in 2037 and hardly any were premeditated" (274). As one of Pete's associates notes,

> In 2024, before the ACIP reshaped human nature itself, things were different. Even if you were an open and optimistic soul, you could never completely trust another person . . . Today fear and suspicion between people are unusual. Improvements in interpersonal communications resulting from the Truth Machine have led to a lower divorce rate, better parenting, better education, and exponential increases in economic prosperity and scientific progress. (348)

Indeed, only ten years after the invention of the Truth Machine, the ubiquity of ACIP scans has rendered the actual machines all but unnecessary: "Few people bothered to look at their ACIP lights anymore. Just knowing the subject was aware of the ACIP was enough to instill confidence" (267).

Yet the ACIP is not foolproof, and its failing eventually threatens to unravel the very fabric of the nations that have incorporated it. When Pete murders Scoggins and reprograms the truth machine to hide his crime, he exposes a fatal flaw in the instrument: its human creator. The defect denotes very clearly that Pete's crime is not simply an individual matter; his ability to fool a machine so integral to the social, national, and cultural fabric of the nation threatens to topple people's faith in the overarching—and quite successful—surveillance system. As the district attorney notes, "*What if everyone now believes that some people can lie with impunity?* . . . Pete's cooperation would be crucial in mitigating the crisis of confidence that would follow the announcement of his confession. People would suspect that the Truth Machine, the most important component of all human interaction, had been rendered useless" (313). That the ACIP's survival and success in the face of failure depends on the public trust should come as no surprise. From Hugo Münsterberg's call for a wider tribunal to the introduction of mechanical mind reading as a literacy, lie detection technologies have always relied on and cultivated a favorable public opinion.

However, as with the other lie detection technologies we have encountered, even progressive visions for social betterment entail trade-offs. "We humans tend to forget that civilization is a system of tradeoffs," notes one of the Supreme Court judges hearing Pete's case. "The perfect world is an unattainable goal. Seldom can members of society gain benefit without exacting a cost, either from themselves or someone else. The human race has opted for survival over privacy, prosperity over individual rights. We have learned that these goals cannot be nurtured simultaneously" (Halperin 1996, 368–69). Ultimately, Pete sacrifices himself to save the system. The courts sentence him to death or treatment for Intermittent Delusionary Disorder (IDD), which involves drug therapy that would reduce Pete's photographic memory to a more average standard. The treatment would allow him to continue some of his work, but hamper his ability to fool any other truth machine. Although Pete appeals the decision, he is eventually forced to choose between the treatment and life imprisonment. He chooses to undergo IDD.

In their punishments for deception, *The Truth Machine* and *The Demolished Man* share one final, crucial quality with each other and with the other lie detection technologies examined in this book: the assumption that physiology and psychology, brain and mind, are bound together in such a way that we can read information from one onto the other. Once apprehended, Ben Reich and Pete Armstrong both undergo procedures that promise to cure them: progressive treatments that make them less threatening to society and spare them from execution. Reich undergoes a procedure called Demolition: "when a man is demolished at Kingston Hospital, his entire psyche is destroyed. The series of osmotic injections begins with the topmost strata of cortical synapses and slowly works its way down, switching off every circuit, extinguishing every memory, destroying every particle of the pattern that has been built up since birth" (Bester 1996, 241). Once again, the brain and mind have been collapsed here: the treatment affects the cortical synapses and in so doing it affects the psyche. And, as in much of the literature (both scientific and literary) I have examined, the true pain, horror, and perhaps even shame of Demolition depend upon a disconnect between mind and body/brain in which the mind is, once again, a helpless observer of its own revelation—or, in this case, destruction. "The horror [of Demolition] lies in the fact that the consciousness is never lost; that as the psyche is wiped out, the mind is aware of its slow, backward death until at last it too disappears and awaits the rebirth" (241).

Reich's treatment is akin to Pete Armstrong's IDD therapy in *The Truth Machine*. Pete is put through a treatment in which he is injected with a serum that alters his brain chemistry. He is assured that the procedure will not alter his identity; "still, he must wonder," notes the narrator, "*Will I emerge intact from the treatment? Or will the person who rises from the chair no longer be me?*" (369). As the narrator continues to speculate about identity, he returns to questions of the biological mind, energy, and psychophysiology that have preoccupied this book in chapters 4 and 5. "Many religions," he argues,

> still contend that each person has an eternal soul. But most scientists will tell you there is no such thing and there is now proof that all conscious thought is simply a form of electrical energy directed into highly complex patterns by the brain. When the brain dies, the electrical energy ceases and there is nothing more. If the scientists are right, Pete will indeed become a different person; the treatment will alter those patterns. (369–70)

The outcome of Pete's treatment is not made clear by the novel. We know he survives, is less intelligent, more caring and calm. We also know that the treatment cures his disability: his ticks, shaking, tongue biting, and "emotional discomfort" (351). At its base, IDD treatment assumes that identity is stored in the gray matter of the brain: "Every human brain has its own wave patterns, as unique as DNA. While the patterns change continuously, certain aspects remain constant, particular to the individual" (Halperin 1996, 201). I took up these issues of brain waves and identity more specifically in chapters 3 and 4; here we are at least privy to the assumption that if one destroys part of the brain, these "individual" waves are disrupted, thrown into new patterns that affect not only the personality but some intangible essence of being. "The set-up is now remarkably similar in appearance," muses the narrator, "to the execution chambers used during the early part of the millennium. . . . Maybe the treatment is symbolic of an execution; when [Pete] leaves the room, a part of him will be dead" (369).[1] Similarly, Ben Reich becomes a "demolished man" whose brain patterns are erased, eliminating his identity so he can be reborn. Evidence for his Demolition is taken not only from his own thoughts, but from several witnesses who see Reich wandering aimlessly around the hospital grounds mumbling incoherently.

Despite their historical separation and their focus on different technologies, *The Demolished Man* and *The Truth Machine* reiterate the patterned history of lie detection, including their assumptions about the measurable effects of lying on the body, assumptions about the body's self-reporting nature, and the definitional entanglement of lies and truth. While both novels assume that science is capable of great feats, including access to the minds of men, they each illustrate the limitations of any knowledge system. As we saw in chapter 4, information is rarely complete or definitive. More often than not, proper disciplinary objects must be selected from an informational chaos. Indeed, Reich's and Pete's strategy is to scramble their pursuers' circuits by overloading them with irrelevant information—distractions in verse form.

My comparison of these two novels, separated as they are by nearly half a century, is not intended to suggest a coherent origin or provide a definitive evolutionary projection for the cultural or scientific success or failure of lie detection technologies. As the epigraphs to this conclusion imply, a genealogical history cannot provide definitive predictions; to do so would be antithetical to the openings that it proposes; but like an important work of science or science fiction, it can speculate about a technology's present by looking to its past and to its potential futures. As William Marston suggested in 1938, "The Lie Detector foundation has been well and soundly laid. From now on its structure of usefulness will rise swiftly in the public view according to the usual and normally predicative course of human events. . . . Do you want to go on, improving the deception test technique and enlarging the scope of its application to human needs? It's up to you!" (1938a, 145–46). And it is also up to us to be cognizant of the assumptions that underlie these technologies, as well as the potential for their abuse.

# Notes

INTRODUCTION

1. The polygraph has enjoyed a long-standing alliance with U.S. government agencies (Ford 2006; Alder 2007). In his 2002 article "A Social History of Untruth: Lie Detection and Trust in Twentieth-Century America," Ken Alder argues that the American fascination with the polygraph is not incidental. Indeed, versions of the machine emerged at a crucial time in U.S. history, just when America was beginning to cope with a mass public and "the rise of new large-scale organizations" (2). The polygraph promised to preserve a spirit of truth and forthrightness in an era of anonymity and potentially compromised national security. In the 1940s Leonarde Keeler, inventor of the portable polygraph, convinced the federal government to use the instrument "to screen security risks," including German POWs who were to be deployed in occupied Germany as police officers (17). By the 1950s, the polygraph was being used by McCarthy's regime to screen for communist sympathizers and homosexuals. Since then, agencies from the CIA to the FBI to local police have employed polygraphs for routine employee screening and suspect interrogation. Despite several blatant failures—as in the Aldrich Ames and Robert Hanssen spy cases—"the instrument is still trotted out as the gold standard in high-profile criminal cases, as a way to plug security leaks, and as an instrument to extract the truth from those suspected of threatening America's safety" (24). For more information on the Lindberg kidnapping case and its intersections with the polygraph, see Jim Fisher, *The Lindberg Case* (1994, 128–29, 181).

2. The ahistoricist impulse I reference here has been analyzed by scholars in several related fields: Lisa Cartwright, writing in a Foucauldian vein, argues that "contemporary scientific imaging [is] a field that persistently has been sealed off from its history (even as it emerges) by a veiling rhetoric of newness and discovery" (1992, 131); I will return to just such rhetoric in chapter 5. Thomas Kuhn (1962) spoke at length about—and other scholars have taken up his call for discussions of—the historicide of scientific knowledge, and the reiteration of humanistic knowledge. In a more general sense, Bruno Latour speaks in terms of the black-boxing of technologies through which the processes of scientific

knowledge production are erased in favor of the final product (*Science in Action,* 1987). Finally, recent work by Robert Proctor and Londa Schiebinger has introduced a new and relevant kind of theoretical analysis: a field called *agnotology,* which can be defined as the cultural study of ignorance. For more information on agnotology, see their edited collection by the same name (2007) and see Proctor's *Cancer Wars* (1996).

3. See chapter 5 for a more complete explanation of blood oxygenation level dependent fuctional magnetic resonance imaging (fMRI).

4. Many of these questions are raised and/or complicated by theory of mind (ToM), a field of study that addresses how we understand other people's consciousness—and the fact that they have consciousness at all. While a full discussion of intersections between deception and ToM is outside the scope of this book, I would point out two central connections: first, that deception research takes for granted that deceptive acts require an understanding and/or assumption of another mind that can be manipulated. And second, that lie detection research assumes that lying is strenuous—that it requires extra work in and of the body—rather than seeing lying as an evolutionarily adaptive strategy. I specifically address the former in the introduction and the latter in chapters 2, 5, and the coda. For more information on ToM, reading the minds of others, and the role of language in the creation of deceptive statements, see Steve Johnson's *Emergence* (2002), Robin Dunbar's *Grooming, Gossip, and the Evolution of Language* (1996), Ronald Schleifer's *Intangible Materialism* (2009), and George Steiner's *After Babel* (1975).

5. There is some disagreement about whether lying and deception are one and the same; see Campbell (2001). While there is a case to be made for lying as a subset of deception (which can encompass a wider range of manipulated information), I will use the terms interchangeably throughout the book. This choice is not out of step with the literature on definitions of deception, the historical literature on lie detection, or the neuroscientific literature on deception and its detection.

6. There is also something to be said about the relationship of fiction to deception, given that fictional literature assumes a certain kind of narrative fabrication. The problem with any simple comparison would be that fictional fabrication does not fit the definition of deception I offered earlier in this introduction; there is no intent to deceive. In fact, the opposite could be said: it is more damning when an author promises nonfiction and delivers a fabrication. In that case, there is an expectation of transparency that is intentionally unmet.

7. For more on emotional inscription, see Dror (1999a, 1999b, 2001a, 2001b); for information on emotional research more generally, see Peter N. Stearns (1994) and Peter N. Stearns and Jan Lewis (1998).

8. The history of lie detection, like the history of scholarship about the polygraph and *the* lie detector, is what we could call *multidisciplinary.* As we will see in chapters 1 and 2, *multidisciplinary* is not an ideal term, given that disciplinary boundaries were not as rigidly defined during the conceptualization, marketing, and deployment of lie detection. Thus, the history of the technologies finds strands in physiology, psychology, philosophy, and law, along with contributions

from police officers, like John Larson and Leonarde Keeler, whom I have already mentioned.

9. See Geoffrey Bunn (1997, 1998, 2007) for a study of the rhetoric surrounding the lie detector.

10. A partial list of scholarship on the history of lie detection includes the following: John Larson, *Lying and Its Detection: A Study of Deception and Deception Tests* (1932); Paul V. Trovillo, "A History of Lie Detection" (1939); Fred Inbau, *Lie Detection and Criminal Interrogation* (1942); L. A. Geddes, "History of the Polygraph, an Instrument for the Detection of Deception" (1974); Geoffrey Bunn, "The Lie Detector, *Wonder Woman,* and Liberty: The Life and Work of William Moulton Marston" (1997); Timothy Hensler, "Comment: A Critical Look at the Admissibility of Polygraph Evidence in the Wake of Daubert: The Lie Detector Fails the Test" (1997); Geoffrey Bunn, "The Hazards of the Will to Truth: A History of the Lie Detector" (1998); David Lykken, *A Tremor in the Blood: The Uses and Abuses of the Lie Detector* (1998); Otniel Dror, "The Scientific Image of Emotion: Experience and Technologies of Inscription" (1999b); Ronald Thomas, *Detective Fiction and the Rise of Forensic Science* (1999); Ken Alder, "A Social History of Untruth: Lie Detection and Trust in Twentieth-Century America" (2002); Ken Alder, *The Lie Detectors: The History of an American Obsession* (2007); Geoffrey Bunn, "Spectacular Science: The Lie Detector's Ambivalent Powers" (2007).

11. I separate literature from fiction here because literature can include myriad textual communications, from (scientific) journal articles, lab reports, technical manuals, Web sites, and media reports to narratives (personal or otherwise), poetry, and fiction. I also do not consider "fiction" to be a mere "fabrication"; instead I see fiction as one more window into the culture that produces the mutual imbrications of literature, science, and technology. That said, fictions have lost and gained cultural cachet depending on their generic positioning. Science fiction and detective fiction were—and sometimes continue to be—ostracized by definitions of "legitimate" literature. In *The Lying Brain,* I assume that each genre has an equal opportunity to teach us something about the larger culture from which literatures, sciences, and technologies emerge.

## CHAPTER 1

1. Hugo Münsterberg was neither the first nor the only scientist to promote the connection between physiology and psychology, the body and the emotions; see Otniel Dror's extensive and excellent work on emotions in the late nineteenth and early twentieth centuries (1998, 1999a, 1999b, 2001a, 2001b).

2. As I will explain in chapter 2, William Marston devised the systolic blood pressure test for deception, what he would later call "the lie detector test" (1938a), an exam that relied on yes/no questions while the subject's blood pressure was recorded by a sphygmograph. By 1938, he is careful to distinguish his test from the lie detector.

3. I have found only two other references to the Luther Trant stories in academic literature: a footnote in Blumenthal (2002) and some brief discussion in Ross and Teghtsoonian (2009).

4. For cultural analyses of lie detection, see Bunn (1997, 2007) and Thomas (1999); for an additional historical perspective, see Alder (2007). In his dissertation (Bunn 1997), the first scholarly history of the lie detector, Geoffrey Bunn argues that the lie detector was created in and by popular culture long before it was ratified or even examined by science; however, he also proposes that these "machines were not . . . lie detectors. They were 'machines for the cure of liars,' 'truth-compelling machines,' or even 'scientific crime detectors' according to *Scientific American,* but they did not detect lies" (Bunn 1997, 43). In his most recent monograph, *The Lie Detectors* (2007), Ken Alder makes a similar argument by claiming that the lie detector was not invented until John Larson, Leonarde Keeler, and William Marston began to experiment with the polygraph: a machine that combined two or more inscription technologies. Ronald Thomas comes closest to recognizing the earlier lie detectors—at least in concept—when he cites Edgar Allen Poe's "The Telltale Heart" and even Daniel DeFoe's claims about criminality and a quickened pulse. But Thomas has little interest in reframing early inscription technologies as lie detectors, and, for that reason, does not recognize *The Achievements of Luther Trant* as a forerunner to the application of physiological instruments and psychological methods to the purpose of lie detection.

5. Pseudoscience shares the problematic associated with "junk" or "bad" science: it is not objective, does not use the scientific method, and/or is not accepted by a mainstream scientific community. For more information on how I use the term *junk science* in this book, please see my explanation in the introductory chapter.

6. Many scholars have chronicled part or all of this historical trend; see, particularly, Hale (1980), Ward (2002), Dror (1999b), and Bunn (1997).

7. For a history of the development of and controversies surrounding the chronoscope, see Schmidgen (2005).

8. Word-association tests typically involved a list of words to which a subject was asked to respond with the first word that came to mind. Each word from the original list acts as a stimulus to produce a response; the time between stimulus and response is said to increase if the subject is trying to conceal information (which in this test translates to a slower response as the subject chooses a nonincriminating word as a response). In the 1870s and 1880s Francis Galton experimented with word-association tests (see "Psychometric Experiments" [1879]), and Wilhelm Wundt also included word-association tests in his experiments (see, e.g., *Grundzfige der physiologischen Psychologie* [1880]). See also Wertheimer and Klein (1904) and Carl Jung (1906); for a meta-analysis see Wertheimer et al. (1992).

9. Münsterberg's catalog of laboratory equipment can be found in his own *Psychological Laboratory of Harvard University* (1893); the chronoscopes appear on pages 12–13.

10. For analyses of the shift in modes of seeing, see Daston and Galison (2007), on objectivity in scientific imaging; Ward (2002), in psychology; Thomas (1999), in early forensics, detective fiction, British/American culture; Dror (1999b), on emotional inscription technologies; Brain (1996), on the shift from

opticism to graphism to digitism; Cartwright (1992), on cinematic language; and Crary (1990), on art history.

11. Bunn, in particular, argues that lies and liars were not disentangled until well into the 1920s, with some remnants of "human kinds" discourse stretching into the 1930s. This is not to say that Münsterberg was uninvested in "human kinds," including the pathological liar, but it is a revision point that is important to my larger arguments in this book.

12. Recognizing lies, instead of analyzing liars, requires a new kind of scientist: the psychological expert who is trained to read the representational system (or language) that these new instruments produce. This gatekeeping discourse can be seen in Münsterberg's description: results need to be interpreted by an expert; each expert and his/her instrument, from the physician to the legal psychologist, has a definitive role.

13. I specify an American context here because applied psychology had already established a relative foothold in France, Germany, and Italy. Part of the American skepticism can be traced to a certain resistance to the transfer of European ideals to an American context. In *On the Witness Stand,* Münsterberg specifies that lawyers "do not wish to see that in this field preëminently applied experimental psychology has made strong strides, led by Binet, Stern, Lipmann, Jung, Wertheimer, Gross, Sommer, Aschaffenburg, and other scholars" (1908, 10–11).

14. Münsterberg references the "new psychology" in "Illusions," *On the Witness Stand* (20), in reference to experimental psychology. He borrows the term from several authors who use it to refer to the boom in psychological laboratories in Germany and the United States, among which Harvard's Psychological Laboratory counts itself the first in the United States. I note this reference because "the new psychology" that we will encounter in chapter 3 was often a reference to a Freudian, psychoanalytic strain of psychology that focused on the unconscious and the hidden self.

15. Münsterberg's ideas, which demanded that he work in the field and account for the effects of the environment on the individual, differentiated him from his German colleagues and his mentor, Wilhelm Wundt, who argued for a structuralist psychology that strove to identify universal laws for human behavior.

16. The "third degree" refers to interrogations that caused physical and psychological pain to the suspect in order to extract a confession. This technique lost prominence in American police practice around the 1930s. See Lassiter (2004), particularly chapter 3, for a more detailed history of the practice in American police interrogations.

17. Charles C. Moore published a diatribe against Münsterberg's ideas, terming them "Yellow Psychology" (1907), and John Henry Wigmore published a similar treatise against Münsterberg's collected essays in 1909.

18. For a wonderful, historical exploration of the struggles between science and the law, see Tal Golan (2004), particularly chapter 6 concerning Hugo Münsterberg.

19. William MacHarg went on to publish with *Colliers* and the *American Post;* his most famous single-authored work is *The Affairs of O'Malley* (1940). Edwin

Balmer published serial fiction with the *Chicago Daily Tribune* between 1916 and 1951; with Philip Wylie, Edwin Balmer published *When Worlds Collide* (1933) and *After Worlds Collide* (1934); the former was made into a film. Between 1927 and 1949, Balmer served as editor of *Redbook Magazine*. Further information about Balmer and MacHarg can be found in several encyclopedias, including Philip Greasley's *Dictionary of Midwestern Literature* (2001), Chris Steinbrunner and Otto Penzler's *The Encyclopedia of Mystery and Detection* (1976/1984), and John Reilly's *Twentieth Century Crime and Mystery Writers* (1985). However, the best sources for information about Balmer are his stories and the press about him available through the *Chicago Daily Tribune* archives.

20. The Luther Trant stories were the forerunners of what would later become known as scientific detective fiction. In this subgenre, the latest technologies are applied to and lauded as important components of police-detective work. The movement also included Cleveland Moffett's *Through the Wall* (1909), Arthur Reeve's *The Silent Bullet* (1912), and R. Austin Freeman's *John Thorndyke's Cases* (1909).

21. I refer here to the eleven stories published between May 1909 and October 1910 in *Hampton's Magazine:* "The Man in the Room" (May 1909), "The Fast Watch" (June 1909), "The Red Dress" (July 1909), "The Private Bank Puzzle" (Aug. 1909), "The Man Higher Up" (Oct. 1909), "The Chalchihuitl Stone" (Nov. 1909), "The Empty Cartridges" (Dec. 1909), "The Axton Letters" (Jan. 1910), "The Eleventh Hour" (Feb. 1910), "The Hammering Man" (May 1910), and "A Matter of Mind Reading" (Oct. 1910). *The Achievements of Luther Trant* (1910) includes the first nine of these eleven stories. Two additional stories, "Decidedly Odd" and "The Day and the Hour," were published in *Top-Notch* in May and July 1915.

22. The collection contains nine of the original eleven stories, published in the same order without any substantive changes to the text. For this reason, and for readability, I typically quote text from the collected edition of the text in this section.

23. *Hampton's Magazine* was the renamed version of *Broadway Magazine* begun in 1898.

24. See Bunn (1997): "In 1925 he had become a 'consulting psychologist'—a new sort of creature who seems to combine the advisory functions of an old-time pastor and country doctor."

25. Note that Balmer and MacHarg (in their foreword) and Luther Trant (in the text) use the same term, "the new psychology," as Hugo Münsterberg.

26. President Hoover convened the Wickersham Commission in 1929 to review the Eighteenth Amendment to the Constitution (which prohibited the production, sale, and distribution of alcohol). The first report from the commission, which issued fourteen reports between 1931 and 1932, was entitled "Lawlessness in Law Enforcement." This report, which stunned the public, revealed many of the unsavory interrogation tactics—including the third degree—used by police. For more information, see Leo (1992).

27. In one story, Trant uses his subject's reactions to a map in order to locate

the scene of a crime. Leonarde Keeler would use a similar technique over thirty years later.

28. Current forensic textbooks (Nickell 1999; Inman and Rudin 2001; Saferson 2007) perpetuate the separation of forensic science and literature by mentioning fiction only to dismiss it. Moreover, these textbooks include fiction from Arthur Conan Doyle, Edgar Allen Poe, and even Mark Twain without a single mention of the Luther Trant stories. That Trant is excised from this group is surprising, but perhaps less than lamentable, for fiction in the context of forensic science's disciplinary history is often derided.

29. See Daston and Galison (1992) for a brief discussion of "imagination" in relation to mechanical objectivity in the later nineteenth and early twentieth centuries: "imagination and judgment were suspect not primarily because they were personal traits, but rather because they were 'unruly' and required discipline" (118).

30. In these stories we also see the origins of the "truth as default" argument so fundamental to later fMRI lie detection experiments: "It is scientifically impossible—as any psychologist will tell you—for a person who associates the first suggested idea in two and one half seconds, like Margaret, to substitute another without almost doubling the time interval" (Balmer and MacHarg 1927c, 51).

31. Marston uses the term *normal* to indicate that there was (and still is) a debate about the efficacy of lie detection when the instruments are applied to pathological liars.

32. See Samuel Walker (1977, 1998). The rhetoric of Luther Trant's arguments for replacing the third degree with psychological investigation continued for many decades. In place of torturous interrogations and subjective accusations, many polygraphists argued, the mechanical detector provided a measure of objectivity and judiciousness to police procedures. As Ernest Burgess argues in his foreword to John Larson's book, "administered by a competent criminological psychologist, it will supersede the third degree and place the interviewing of suspected individuals upon both a scientific and a human basis" (1932, xi). Even Fred Inbau's *Lie Detection and Criminal Interrogation* (1942) argues that "the use of a lie-detector does not constitute a 'third degree' practice. The temporary discomfort produced by the blood pressure cuff is too slight to warrant objection, and the test procedure is of such a nature that it is extremely improbable that it would encourage or compel a person to confess a crime which he did not commit" (68). Here, and elsewhere in the detection deception literature, lie detection allows police to gain their desired confession from most guilty individuals through the strangely sanitary and uncontaminated psychological pressure of an "objective" machine. Criminals incriminate themselves, while innocent suspects are (rather miraculously) unafraid of the resulting graphic evidence.

33. See Ward (2002, 119–20) for history of these instruments.

34. See Walker (1998).

35. See Walker (1998): "Police chiefs enthusiastically embraced the new technology, believing it would make possible efficient and effective police service" (166).

36. Marston was a psychologist and lawyer; Larson, Keeler, and Vollmer were police officers; later, Fred Inbau, a lawyer, became one of the most outspoken and well-respected polygraphers.

37. It should be noted that Marsten's research began in 1913. He does share credit with Vittorio Benussi for the application of instrumentation to lie detection during the 1910s. Marston worked on blood pressure, while Benussi worked with respiration. For more information about Benussi, see Stucchi (1996).

38. Vollmer was the police chief at Berkeley; for more information about Vollmer's life and career, see Gene Carte and Elaine Carte's *Police Reform in the United States: The Era of August Vollmer, 1905–1932* (1975); see also Robinson, Carte, and Carte, *August Vollmer: Pioneer in Police Professionalism,* vols. 1 (1972) and 2 (1983).

39. More information on John Larson and Leonarde Keeler, in particular, can be found in Ken Alder's book *The Lie Detectors* (2007).

40. Keeler's "norm" tests include his "card test" and "peak of tension test"; I discuss the former in chapter 2.

41. See particularly Bruno Latour's *Science in Action: How to Follow Scientists and Engineers through Society* (1987), chapter 5, "Tribunals of Reason," for an explanation of how networks affect "allies" within and in conjunction with scientific fact production and distribution.

42. Importantly, Marston actually goes to great pains to deny that he is the sole inventor of the lie detector; he does, however, claim to have invented the lie detector *test* using a sphygmomanometer.

43. In 1993, the *Daubert v. Merrell Dow* case challenged the standard of evidence introduced by *Frye v. U.S.*; instead of relying on "general acceptance" by a community of scientific experts, *Daubert* gives control to the judiciary: "The Court adopted a new framework for evaluating the reliability of scientific evidence, based on four considerations: falsifiability, peer review, error rates, and 'acceptability' in the relevant scientific community" (Cheng and Yoon 2005, 477). *Daubert* holds sway in federal courts, but because it was not a constitutional case, individual states can decide which standard to uphold. Some states still rely on the *Frye* standard, while others have adopted the *Daubert* standard. For more information about states' acceptance and the value of expert testimony, see Cheng and Yoon (2005), Hamilton (1998), and Kauffman (2001).

44. See also H. Cairns, *Law and the Social Sciences* (1935); Edward Robinson, *Law and the Lawyers* (1935); Jerome Frank, *Law and the Modern Mind* (1930); and D. McCarty, *Psychology for the Lawyer* (1929). For contemporary arguments, see Steven Goldberg, *Culture Clash: Law and Science in America* (1994), and Robert Smith and Brian Wynne, *Expert Evidence: Interpreting Science and the Law* (1989).

45. For more information on Gernsback as an editor of science fiction and other pulp fiction, see Mike Ashley, *The Gernsback Days* (2004); see also Siegel (1988).

46. This phrase is taken from Gernsback's first editorial for *Amazing Stories* (April 1926).

47. He would eventually publish new material in *Amazing Stories,* but

throughout his tenure as editor (until he was ousted from the position due to bankruptcy in 1929) he continued to republish stories.

48. The stories published in *Amazing Stories* are as follows: "The Man Higher Up" (Dec. 1926), "The Eleventh Hour" (Feb. 1927), "The Hammering Man" (Mar. 1927), and "The Man in the Room" (Apr. 1927). At least five more (with some duplication) were republished again in *Scientific Detective Monthly* as follows: "The Fast Watch" (Jan. 1930), "The Man Higher Up" (Feb. 1930; 1:2), "The Man in the Room" (Mar. 1930; 1:3), "The Hammering Man" (Apr. 1930; 1:4), "The Eleventh Hour" (May 1930; 1:5). The lie detector was also featured on the cover of the Nov. 1929 issue of *Scientific Detective Monthly* that featured "Science, the Police—and the Criminal," an article by Ashur van A. Sommers (Apr. 1930, *Scientific Detective Monthly*).

49. Arthur Reeve, who made a name for himself in American scientific detective fiction as the author of the John Thorndyke stories, was editorial commissioner for the magazine.

50. "Psychic evidence" refers to evidence derived from lie detection, not to parapsychology, but the use of this derogatory descriptor revels some of the struggle to separate psychology from the parasciences.

## CHAPTER 2

1. Baby parties were often put on by sororities at the universities where Marston was employed; they involved diaper-clad pledges being bound by their sisters and spanked as a hazing ritual. Marston, who was interested in theories of submission and dominance—which would later manifest themselves in his DISC personality theory—saw such parties as a significant example of the basic human tendency to enjoy submitting, particularly to powerful women. In this same vein, Marston created Wonder Woman as a role model for little girls: noting that "not even girls want to be girls so long as our feminine archetype lacks force, strength, and power" (1943, 42–43). Finally, Marston's biography of Julius Caesar is far from standard fare, being a largely pornographic tale of domination and submission.

2. Marston's trendy sense was correct: between 1914 and 1915, as he pursued research for his PhD in psychology, another researcher, Vittorio Benussi, published a study concerning the correlation between respiration and deception. Although Marston acknowledged Benussi as a forerunner, he also took issue with several of Benussi's assumptions, methods, and results. Both of Marston's initial studies were responses to Benussi's 1914 experiments on respiration and deception. There was a particular rivalry between the two men regarding credit for the first lie detection experiments.

3. "Systolic Blood Pressure Symptoms of Deception" (1917) details William Marston's experiments at the Harvard Psychology Laboratory performed between 1914 and 1915.

4. In his 1938 book *The Lie Detector Test* about its potential and pragmatic use, Marston publicly touted lie detection as a "psychological medicine" capable

of curing various emotional, social, and political ills (1938a, 15). He planned, ultimately, to apply his findings to other field situations, including criminal interrogations, relationship counseling, and psychological reeducation programs.
See Geoffrey Bunn (1997, 1998), Rhodes (2000), Alder (2007), and Lykken
(1998) for some coverage of Marston's field applications of lie detection and the
deceptive consciousness.

5. These terms are entirely mine and represent an attempt at a heuristic that
binds and describes the kinds of experimental protocol used by Marston. i would
stress here that Marston, like Münsterberg before him, was not interested in the
liar as type. I say this for two reasons: first, Marston resisted later polygraphers
who insisted on including "typed" criminal graphs in their books/manuals for lie
detection; second, Marston did not record systolic blood pressure during physically criminal activities (like the mock crimes) but afterward during deception.

6. Lest these dates seem confusing, please note that Marston's research was
conducted between 1913 and 1922. He published academic papers concerning
this research between 1917 and 1927.

7. For a wonderful analysis of "objectivity," including its shifting definition
and imbrications with morality during the nineteenth century, see Daston and
Galison (1992, 2007). For discussions of witnessing in laboratory experiments,
see Shapin and Schaffer (1985) and Haraway (1997); Haraway also discussed
the problem of objectivity as a "view from nowhere" in *Simians, Cyborgs, and
Women* (1991).

8. From Claude Bernard's experiments of the 1860s, in which he attempted
to link emotions to responses in the heart, to the work of Josiah Stickney Lom-
bard, Angelo Mosso, Alfred Binet, and Charles Féré, emotions became objects of
scientific inquiry (Dror 2001b).

9. Although Marston inaugurated many traditions for lie detection, he was
markedly opposed to the inclusion of graphs in his later work *The Lie Detector Test*
(1938a). In fact, he was faulted for the lack of graphs by the likes of Fred Inbau
and others. With the exception of Marston, the burgeoning field of polygraphy
took to graphic representations as a means to categorize and graphic records according to not only type of lie but type of criminal. In his introductory statement
to John Larson's book *Lying and Its Detection* (1932), August Vollmer proudly
notes that "in this book the vast literature relating to deception is critically analyzed. . . . The various types of deceptions are defined and classified. Deceivers
are placed in their pigeonholes so that each type is readily recognizable" (vii).
Both Larson's and Inbau's books on lie detection include multiple graphs that
are clearly labeled and categorized. Most are similarly titled: "Record of an Arson
Suspect," "Record of a Bomber," "Record of a homosexual suspect lying about
his guilt" as if to suggest that these are templates or standard records to be used
for comparison. Within the scientific community, these records produce and are
produced by a false sense of control. If criminals could be "pigeonholed," they
were not only tractable but knowable. In fact, lie detection allowed for the creation of new criminal categories. Even the most intangible (and immoral) "criminal" activities—such as those perpetrated by the "homosexual suspect" who lies
about his "guilt"—could be identified.

10. For a history of mechanized "objectivity" and its relation to graphic technologies, see Daston and Galison (1992), particularly 115–17.

11. The pagination of Marston's dissertation is such that each chapter begins on page 1, despite the fact that the table of contents uses continuous pagination. This quote is taken from page 3 of chapter 5: "Systolic Blood Pressure Symptoms of Deception."

12. Later, more familiar versions of "the lie detector" or polygraph included multiple measures, including systolic blood pressure, respiration, galvanic skin conductance, and heart rate/pulse. Marston was one of the first proponents of these combination tests; however, in the experiments discussed here, he was devoted to the systolic blood pressure as a primary test.

13. There was some debate later, particularly from John Larson, as to Marston's technique. While Marston used the Tykos sphygmomanometer, he used the palpitatory rather than the ascultatory method. The latter requires a stethoscope rather than the hands of the examiner. In his "Modification of the Marston Deception Test" (1921), John Larson argues for the ascultatory method using similar logic to that which we have seen thus far: as "it is desirable to eliminate all personal factors whenever possible" (392), mechanical recordings are more objective.

14. See previous note concerning the pagination of Marston's dissertation. This quote was taken from the "Summary" section of Marston's dissertation, page 4.

15. For more information about the ways in which modernity fragmented, measured, and monitored the body, see my essay "Matter for Thought: The Psychon in Neurology, Psychology, and American Culture, 1927–1943" (2010); see also chapter 3 of this book and Armstrong (1998), Rabinbach (1990), and Seltzer (1992).

16. Marston would later advocate for the lie detection test to include multiple measures, including respiration, blood pressure, and heart rate; see *The Lie Detector Test* (1938).

17. Although he does not provide a sample of any questionnaire or materials used by the investigator, it appears that in most postcrime introspection, subjects chose from a catalog of possible feelings. Indeed, different subjects often use the same terms to describe their individual feelings. If Marston did provide a list of possible descriptors, an argument for the scripted performance of lie detection exams gains even more credence.

18. See the following by William Marston: "Systolic Blood Pressure Symptoms of Deception" (1917); "Reaction-Time Symptoms of Deception" (1920); "Psychological Possibilities in the Deception Tests" (1921a); "Sex Characteristics of Systolic Blood Pressure Behavior" (1923); "A Theory of Emotions and Affection Based upon Systolic Blood Pressure Studies" (1924); "Negative Type Reaction-Time Symptoms of Deception" (1925).

19. Fear and rage, along with pain, were initially identified by Walter Cannon in *Bodily Changes in Pain, Hunger, Fear, and Rage* (1915) as emotions that specifically affect blood pressure. Marston rules out pain, noting that according to Vittorio Benussi and the vivisectionists "only the diastolic pressure is

significantly altered by pain" (1917, 121); see also Benussi (1914).

20. Marston argues in 1924 that there are two types of deception: negative and positive. His views are challenged by a contemporary, Eva Goldstein, who argues that Marston's experimental setup was mechanical and did not necessarily get at "deception." Marston's response is that the negative type can be explained: the subject simply had no consciousness of deception.

21. I use the gender neutral pronoun here because Marston's experiments did involve both women and men. Indeed, his work with women and his concepts of gender and feminism have been addressed by scholars (Bunn 1997; Rhodes 2000). Had I the space and time, I would elaborate further on Marston's theories about gender, but suffice it to say: another book could be written about Marston and the female, femininity, and feminism.

22. For more information on the council, see Robert M. Yerkes, "Psychological Work of the National Research Council" (1923). He explains that "the National Research Council was organized in 1916 under the auspices of the National Academy of Sciences and the Engineering Foundation, primarily to place the scientific resources of the country at the command of the federal Government" (172).

23. Only after the failures of the *Frye* case did Marston turn away from laboratory research on the deceptive consciousness and toward more general research on emotion and multiple popular venues in search of a public following. For the former, see the following by William Marston: "A Theory of Emotions and Affection Based upon Systolic Blood Pressure Studies" (1924); "Studies in Testimony" (1924–25); "Negative Type Reaction-Time Symptoms of Deception" (1925); "Motor Consciousness as a Basis for Emotion" (1927); *Emotions of Normal People* (1928); "Bodily Symptoms of Elementary Emotions" (1929). For the latter, see "Blondes lose out in film love test" ("Blondes" 1928); "New Facts about Shaving Revealed by Lie Detector!" (Marston 1938b); and *The Lie Detector Test* (1938a).

24. Marston was not the first to introduce lying to the laboratory; see Jung (1906); Wertheimer and Klein (1904); Yerkes and Berry (1909); Henke and Eddy (1909). These experiments and their relation to crime, specifically, are cited in Eva Goldstein's (1923) experimental response to William Marston's early work.

25. The different types of stimuli can be seen in the following experiments: refusing orders (using cards) happened at the Harvard Psychological Laboratory between 1913 and 1914; creating alibis and the suppression of truth occurred at the Harvard Psychological Laboratory in 1915; mock crimes were used in the Harvard Psychological Laboratory for the 1915 experiments as well as those at Camp Greenleaf in 1918 (published results appear in his 1921a); occlusion of information (involving actual criminal testimony) can be found in Marston's experiments conducted in 1920 (see Marston 1921a, 566).

26. Nearly all lie detection subjects are asked to play the role of the antagonist (at one time or another) given the ways in which "norms" are generally established in lie detection experiments. In the course of an experiment or examination, another kind of norm is also established when subjects are asked to sit

quietly at the beginning and end to establish their baseline blood pressure, respiration, heart rate, etc. This measure is one kind of "norm": a subject at rest, neither actively lying nor telling the truth. It is often used as a baseline for later "truthful" responses, which are expected to register less intensely on the graphic record.

27. In some cases this experiment uses cards that have two columns of numbers. The same basic procedure is used except that subjects are (also) asked to add or subtract the numbers in addition to moving from top to bottom and left to right.

28. See, for example, a sampling of the latest articles on brain-based detection: Cutmore et al. (2008); Vandenbosch et al. (2008); Ambach et al. (2008); Lui and Rosenfeld (2008); and Lui and Rosenfeld (2008).

29. In a very literal way, the laboratory and field are incommensurable: In detailing a future course for his and others' experiments on deception, Marston explains that "the examination should be private, with carefully controlled conditions; and means at hand for recording involuntary movements, muscular contractions, and sudden or suppressed laughter" (1917, 7). Such carefully monitored conditions are often not present in the field.

30. These subjects are suspected criminals who are given a lie detection exam the results of which are verified by conviction or exoneration in court or by a medical expert.

31. The suspects involved in Marston's experiments are given a lie detection test before undergoing a medical and juridical examination that results in a legal verdict of guilt or innocence.

32. Introspection can occur at any point during a psychological experiment but is usually requested after a given stimulus. During Marston's experiments, subjects were expected to offer introspective comments not about their mental processes but about their emotional reactions after the mock crime and lie detection exam had ended.

33. Leonarde Keeler's card trick functions in a similar way, though it was more frequently used to establish a norm before a longer lie detection examination that might include word association or other tests. At the beginning of an exam, the antagonist is asked to choose a playing card from a pile of ten and to remember the number and suit. The cards are shuffled and the experimenter instructs the antagonist to respond negatively to the question, "Is this the card you chose?" even when his card is displayed. According to Keeler's own experiment, "under these conditions the subject tells one lie, and without fear or anger or other apparent emotional stress. Seventy-five subjects were tested, of which only four showed no response to indicate the chosen card" (Keeler 1930, 43).

CHAPTER 3

1. When I use the term *visible*, I do not mean to imply that these objects preceded their visualization—and construction—as proper objects of scientific disciplines; I use the term because, in many cases, the "discovery" was predicated on being able to *see* an object through the mediating lens of a machine.

2. By including both verbs, *to see* and *to name,* I wish to draw attention to the liminality of these objects: they are not simply discovered outside of language or ideology; they are both material and theoretical.

3. For more information on theories of energy, force, thermodynamics, and matter, see Luckhurst (2002); Clarke and Henderson (2002); Smith (1998).

4. Being able to see into the body and brain to fragment them into various components fit well with the modern "desire to *intervene* in the body; to render it part of modernity by techniques which may be biological, mechanical, or behavioral" (Armstrong 1998, 6). The modern individual was splintered into various objective, quantifiable, and controllable components that could be monitored and modified by specifically developed technologies. Under this rubric, diagnoses could be made and treatments offered. Thus, modernity "brings both a fragmentation and augmentation of the body in relation to technology; it offers the body as lack, at the same time as it offers technological compensation" (3). The newly modified body would be one that does not suffer from the crises associated with modernization and modern civilization such as the threat of irrationality and fatigue. This ideal body would be perfectly rational and self-controlled.

5. Telepathy has its roots in the nineteenth-century work of Frederick Myers, who coined the term in 1882. Its literal Greek translation is "distant (*tele*) intimacy or touch (*pathos*)" or "intimate distance" (Luckhurst 2002, 1, 3).

6. J. B. Rhine renamed telepathy and other psychic forces like telekinesis as forms of extrasensory perception (ESP). He also brought telepathy into the laboratory at Duke, through the use of Zener cards that are marked with cross, square, circle, star, or wave. For more information, see Luckhurst (2002, 252–53).

7. *Mind to Mind* is the collected combination of *La Telepathy* (1921), articles from *Revue Metapsychique,* and miscellaneous materials from 1935.

8. Of course, scientific representations are equally symbolic and in need of subjective/human interpretation. I do not mean to binarize the two types of writing in this case; but I am interested in the specific types of symbolism that continually resurface in telepathic accounts.

9. I should note here that references to telepathy's symbolism—particularly its primitivism—resonate with both psychoanalytic theory and Jungian psychology. The latter provides a better analytic tool, primarily because there was a large divergence between parapsychologists and psychoanalysts concerning the unconscious. According to Jan Ehrenwald's *Telepathy and Medical Psychology* (1948), "psychoanalysts relegate the claim of telepathy into the realm of phantasy, of wishful thinking governed by the pleasure-principle, by the craving for what Freud calls omnipotence of thoughts, common to children, neurotics, and primitive men" (43). Ehrenwald also takes issue with Freud's suggestions that telepathic transmissions are merely the result of repressed passions and emotions (46–47). What becomes confusing is the language used by parapsychological researchers. Warcollier argues, for example, that the unconscious is the best transmitter of telepathic impulses: "in the unconscious, thought is dynamic. . . . when it becomes conscious, it loses much of its energy" (1963, 63). However, he is re-

ferring to Frederick Myers's, not Freud's, conception of the "subliminal self"—the parapsychological version of the unconscious. The main difference between the two versions of the unconscious is accessibility. Myers argued that the unconscious was simply outside normal human perception, much like certain types of light in the spectrum, and one could be trained to access telepathy just as we learn to use our five senses. "For Freud, on the other hand, the unconscious falls outside human self-perception because it is *essentially* unconscious" (Keeley 2001, 784).

10. In one laboratory experiment, W. H. Carrington used a large collection of drawings from which the telepathic agent could choose.

11. The history of Berger's experiments can be found in narratives by Peter Gloor (1994) and David Millett (2001). In addition, Gloor has published the only English translation of Berger's first twelve publications on EEG (Gloor 1969).

12. See E. D. Adrian and B. H. C. Matthews (1934), "The interpretation of potential waves in the cortex."

13. In his article on Hans Berger's scientific career, David Millett explains one possible inspiration for EEG. Before developing the now ubiquitous technology, a nineteen-year-old Berger enlisted for a year of military service in Würzburg, Germany. The year was 1892; Berger was avoiding both the fast pace of life in Berlin and his father's desire that he pursue medicine. Yet, his short stint in the military would, in fact, inspire him to study the psychophysiological connections between mind and brain. "One spring morning, while mounted on horseback and pulling heavy artillery for a military training exercise, Berger's horse suddenly reared, throwing the young man to the ground on a narrow bank just in front of the wheel of an artillery gun. The horse-drawn battery stopped at the last second, and Berger escaped certain death with no more than a bad fright. That same evening, he received a telegram from his father, inquiring about his son's well being. Berger later learned that his older sister in Coburg was overwhelmed by an ominous feeling on the morning of the accident and she had urged their father to contact young Hans, convinced that something terrible had happened to him. He had never before received a telegram from his family, and Berger struggled to understand this incredible coincidence based on principles of natural science. There seemed to be no escaping the conclusion that Berger's intense feelings of terror had assumed a physical form and reached his sister several hundred miles away." (Millett 2001, 524)

14. Berger's original German: "Möge es mir gelingen, den schon seit über 20 Jahren gehegten Plan zu verwirklichen u. so eben doch eine Art Hirnspiegel: das Electrenkephalogramm zu schaffen!" Diary, November 16, 1924 (quoted in Borck 2001, 583).

15. Indeed, subsequent studies, instrumental recordings, and literary accounts of brain activity measurement continue to address and buttress Berger's paradoxical questions; see Obermann (1939).

16. *Science Fiction Series* was a series of booklets that were advertised for ten cents each in one of the premier science fiction magazines, *Wonder Stories*.

17. We could look, for example, to the invention of the psychic police

officer, examples of which include J. U. Giesy's Semi-Dual (1912), a psychologist who can detect crime through psychic vibrations; Jules de Grandin (1925); and in 1935 Doctor Occult. All three detectives battle the evils of the supernatural world (whose demons often appear in animalistic form) using various thought detectors that can measure the evil in thought energy.

18. Examples include Louis E. Bisch's *Your Inner Self* (1922) and James Oppenheim's *Your Hidden Powers* (1923).

19. This was in contrast to the behaviorists who believed that "instinct was largely discarded and man's behavior construed to be the result of accidental associations of primitive visceral and behavioral reactions" (Burnham 1988, 83).

20. Even the Occult detectives (Semi-Dual, Jules de Grandin, and Dr. Occult) seek social justice through the application of their powers. See, for example, Guisey's the "Ivory Pipe" in which Semi-Dual is described as a "student of the forces of universal vibration, of those things commonly called 'occult' quite appropriately, too, since occult means really only 'hidden' from the average mind" (12). In Guisey's "The Unknown Quantity," we are told at one point that "the hour approaches when all things concerning this matter shall be revealed and no longer lie hidden—when the mask of secrecy and cowardly hiding shall be rent in twain—when the hand—the right hand—which struck death to a worthy soul from behind shall be made known, and its selfish motive laid bare—when the unknown shall become at last the known" (16).

21. The novel was originally a serial published by *Galaxy* in 1951. Although the novel may sound like a riff on Philip K. Dick's "Minority Report," Bester's novel was published several years before Dick's version of precognitive police surveillance.

22. The Control Question Test (CQT) was developed by John Reid during the 1940s, and his first publication about the technique came out in 1947, followed by a second, revised version of the test in 1955. The test is similar to—and indeed, the precursor to—what would become the Guilty Knowledge Test (GKT), developed by David Lykken in the 1980s and, later, Brain Fingerprinting itself.

23. See, for example, T. Proctor Hall's "Dr O'Glee's Experiments," *Amazing Stories Quarterly* (1929), in which a ray cures criminality; John Campbell's "Solarite," *Amazing Stories* (1930), in which kleptomania is cured by a ray treatment; or Harl Vincent's "Barton's Island," *Amazing Stories* (1929), in which, again, a ray cures criminality.

24. Bester's Espers are divided into three classes: 1st, 2nd, and 3rd. The 3rd-class Espers can only "peep" the conscious level of thought, 2nd-class Espers can "peep" the preconscious, and 1st-class Espers can "peep" the unconscious (see 13–14). Another bit of useful but tangential information: Espers belong to a Guild, pledge an Esper Oath (akin to the Hippocratic oath), and live in a fairly tight-knit community: the punishment for misbehavior is ostracism. There is also a eugenic undertone to the Esper society. The Guild requires that they intermarry (Espers cannot marry normals). As for the "mind block," see my conclusion on pattern recognition.

25. Indeed, the Espers are a fickle and cliquish crowd who actively judge and even exile members of their own clan based on how they think. When a 3rd-class Esper is brought as a date to the party, she is scolded by her boyfriend (a 2nd-class Esper) for using words: "'Don't talk,' the lawyer shot at her. 'This isn't a 3rd Class brawl, I told you not to use words'" (31). Here, as in "The Thought Translator," words are neither necessary nor desirable; and, like the earlier narrative, the truth about a person is not what is said but what is in their mind. In this case, the lawyer's fiancée is deemed "naïve and not deeply responsive. Obviously a 3rd" (Bester 1996, 31) because her thought pattern ("TP") is so unsophisticated. Another Esper, who figures prominently in the story for his exile from the community, loses his identity when he is no longer allowed to participate in Esper conversations. When we first meet Jerry he is standing outside the light and warmth of this very same party, "huddled in the ♯ shadow of the limestone arch . . . He was cold, silent, immobile, and starved. He was an Esper and starved . . . living on a sub-marginal diet of words for the past ten years, was starved for his own people—for the Esper world he had lost" (33).

26. Within Bester's narrative, the Espers produce photograph-like evidence, but they do not use instruments to do so; in this respect they represent a divergence from Münsterberg's instrumental mental microscope.

CHAPTER 4

*Body snatching* generally refers to the crime of grave robbing that frequently occurred in the nineteenth century as medical students required bodies for dissection. An exemplary literary example is Robert Louis Stevenson's narrative ghost story "The Body Snatchers" (1884, republished 1906). Due to religious constraints, few bodies were available for experimental use. While criminals' bodies were often used, because they were easier to acquire, the shortage of bodies created a market for grave robbers.

1. Simon Cole discussed Sir Francis Galton's obsession with character in *Suspect Identities* (2001); he also cites an excellent essay by Paul Rabinow, "Galton's Regret: Of Types and Individuals" (1992).

2. For another discussion of the constitutive power of classification, see Benjamin Cohen's article "The Elements of the Table: Visual Discourse and the Periodic Representation of Chemical Classification" (2004).

3. One fascinating aspect of fingerprinting history is that it is continually retold in popular and academic contexts, yet rarely will one author acknowledge another. In forensics, see Nickell and Fischer (1999); in literature, see Thomas (1999); in history, see Cole (2001), Sengoopta (2003), and Browne and Brock (1953).

4. See Ronald Thomas's *Detective Fiction and the Rise of Forensic Science* (1999), Stephen Gould's *The Mismeasure of Man* (1996), and Simon Cole's *Suspect Identities* (2001).

5. Among Galton's contributions was the classification of fingerprints into three main taxonomic categories: whorls, arches, and loops.

6. Simon Cole notes that before *fingerprint* was coined, terms such as *digital*

*signatures* or *digital photographs* were used (2001, 250). The double valence of *digital* (relating to fingers and information) provides a telling connection between originary systems of biometric identification and newer, ironically disembodied versions of the same data.

7. The controversy over who discovered fingerprinting is quite complex and involves other issues: Who is granted credit for actually realizing the implications of digital marks? Who first recognized their use value? Who first created a system to manage and classify prints? Herschel's work in India was adopted, further theorized, and used by Sir Frances Galton and Edward Henry, for example. Others were also working independently on fingerprinting: Dr. Nehemiah Grew presented work on fingerprints to the Royal Society in 1684; Johannes E. Purkenje first classified fingerprints in his 1823 thesis; Thomas Bewick included an engraving of his thumbprint in an illustrated edition of *Aesop's Fables* in 1818; Henry Faulds, a Scottish physician, was the first to publish his work with fingerprints in Tokyo (his letter on the subject appeared in *Nature* on October 28, 1880).

8. For a history of Bertillon Anthropometry, see Cole (2001), particularly chapter 2.

9. I specifically refer here to the collection of fingerprints from bodies, not crime scenes. The retrieval of prints (partial and otherwise) from the scene of a crime often requires great skill. See J. Nickell and Fischer (1999) for a more specific explanation.

10. The most telling example of the politics of criminology and criminalistic techniques like fingerprinting comes from Havelock Ellis's *The Criminal* (1890, republished 1913), a volume in which he calls for a British science of criminology. Although Ellis's work stems largely from Cesare Lombroso's *L'Homme Criminel* (1887, published in English in 1911), he argues that his own science will not seek to collapse criminal science with politics (Thomas 1999, 207). He claims that Lombroso's theories of atavism may not be objective enough. Instead, he offers up hundreds of criminal measurements and argues for their facticity. However, as Ronald Thomas points out, more often than not, the examples used by Ellis illustrate similarities between criminals and "the lower races" or "the lower apes" (1999, 208).

11. Only after fingerprinting was proven successful in India and other colonies was the technology implemented in European criminal investigation.

12. Herschel does note that he first proposed the fingerprint as a means to identify criminals during their trials in 1877 (1916, 22–24).

13. The Bureau of Identification became the Federal Bureau of Investigation (FBI) in 1934. For a history of the rise of the FBI and J. Edgar Hoover, see Breuer (1995). For more information on the New Deal politics that led to the rise of a centralized police state (including the FBI), see Potter (1998).

14. To promote the campaign and the FBI, Hoover initiated a tour during which he spoke to various groups around the country. Although an analysis of these pamphlets is beyond the scope of this chapter, interested readers should look to the following addresses made by Hoover: "Modern Problems of Law Enforcement" (1935a); "Local Law Enforcement in Relation to National Crime" (1936b); "The Influence of Crime on the American Home" (1936a); "Patriotism

and the War Against Crime" (1936c); "Science in Law Enforcement" (1936d); "Progress in Crime Control" (1938).

15. Hoover worked in conjunction with August Vollmer, who was famous for his work to reform the police. Vollmer, who began his career in Berkeley, eventually moved to Chicago. He was, as we saw in chapter 1, involved in the development of lie detection. Indeed, as Ken Alder has argued, Vollmer was a central character in lie detection debates between his protégés, John Larson and Leonarde Keeler. For a complete bibliography of Vollmer, see *Police Reform in the United States: The Era of August Vollmer, 1905–1932,* by Gene E. Carte and Elaine H. Carte (1975). For Hoover's own explanation of the modern crime lab, see "Scientific Methods of Crime Detection in the Judicial Process" (1935b).

16. Hoover's deployment of fingerprinting continued through the second Red Scare of 1947, through the 1950s and into the 1960s with little alteration. Between the 1960s and 1970s, a second wave of police reform corresponded with the civil rights movement; see Walker (1998).

17. I should note that when Finney republished "Body Snatchers" as the novel *Invasion of the Body Snatchers,* he substantially revised the FBI's role in the final scenes of the story, leaving Miles and Becky, the protagonists, alone in their fight against the invasion. The FBI never swoops in to rescue them from the fields of battle. The defeat of the FBI reminds us of the aliens' power. If the pod-people allegorically represent the *indistinguishable* Other (the communist, foreigner, and immigrants) who can infiltrate unnoticed, they are also representative of the *unmarked* colonizer. In the penultimate scene of the novel, for example, as Miles and Becky set fire to a field of alien seed-pods and watch as the invaders retreat into space, Miles figures himself (and humanity) as the potentially subjugated race, as those who would not be colonized: the aliens "could tell with certainty that this planet, this little race, would never receive them, would never yield" (Finney 1996, 214). He even repeats Churchill's wartime speech as they ruin the pods in the field: "*We shall fight them in the fields, and in the streets, we shall fight in the hills; we shall never surrender*" (214). The speech, "We Shall Fight Them on the Beaches," given on June 4, 1940, is a response to the potential German invasion of Britain during World War II. In this context, Miles's comment resonates more closely to fears of subjugation than amalgamation.

18. *Invasion of the Body Snatchers* was later republished to coincide with the 1978 movie adaptation. In this later version, a few minor changes are made, along with the major change of the setting's date: 1976.

19. I am specifically referring to the novel here because this chapter will not address the four major film adaptations of *Invasion of the Body Snatchers.* Ironically, the novel is often forgotten, rather than being foundational to criticism of this alien invasion narrative. Here, I want to foreground the novel and reposition *Invasion of the Body Snatchers* as a story about biometrics given the cultural context of J. Edgar Hoover, the FBI, and fingerprinting.

20. "The Body Snatchers" first appeared in the November and December 1954 issues of *Colliers* as a serial, before appearing as a novel in 1955. It has since been adapted as a film four times: Don Siegel's 1954 *Invasion of the Body Snatchers*, Phillip Kaufman's 1978 *Invasion of the Body Snatchers*, Abel Ferrara's 1994 *Body Snatchers*, and, most recently, Oliver Hirschbiegel's *The Invasion* (2007).

21. This trend to see identity as central to national security is seen in other science fiction of the era: Philip K. Dick's *Do Androids Dream of Electric Sheep?* (1968, republished 1996), "Imposter" (1953, republished 2002), *The Simulacra* (1964, republished 2002), and Robert Heinlein's *The Puppet Masters* (1951, republished 1986) are all fine examples.

22. It is certainly true that *Invasion of the Body Snatchers* is set in California, a liminal space in which races, cultures, and people ostensibly meet and mix. "In the post-war era and beyond," argues Katrina Mann, "California has been a particularly loaded cultural sign with regard to race relations and white hegemony. As a popular site for Asian and Mexican immigration and for Mexican American, Asian American, and African American migration, California has represented an unparalleled level of racial 'amalgamation'—a vision, sometimes sanguine, mostly garish, of America's racial future in the age of *Brown v. Board of Education*" (2004, 54). Mann, like others before her, argues that *Invasion of the Body Snatchers* "articulated a speculative and hyperbolic confrontation between suburban whites and invasive alien others in terms that reverberated with familiar postwar tropes about racial integration" (55–56).

23. For fingerprint references see 85, 86, 88, and 92 in the 1996 edition of the 1955 *Invasion of the Body Snatchers* text.

24. Here, it is important to note a crucial reversal: the pod-people's unmarked character is unnerving not because they are an unidentified subjugated race but because they are the colonizers of the novel come to assimilate the humans of Santa Mira, California. Thus, even in Finney's 1950s novel, biometric identification is embedded in colonial issues of power, marked bodies, and systems of biological identity that began centuries before.

25. DNA's function was first explained in detail by Watson and Crick in a 1953 *Nature* article and subsequently taken up by the popular press beginning in 1953, only a year before *Body Snatchers* was initially published in *Colliers*.

26. Friedrich Miescher worked on the nuclei of white blood cells while at the University of Tübingen.

27. "What nobody understood before the Cavendish Laboratory men considered the problem was how the molecules were grooved into each other like the strands of a wire hawster so they were able to pull inherited characters over from one generation to another" ("Clue" 1953, 17). Thanks to Watson and Crick, science and society learned that DNA "controls the production in the living body of animal and plant of other molecules, such as proteins and enzymes, with specific properties" (Laurence 1954, 31). Finney most likely read these reports in the *New York Times* given that one of his characters, the writer Jack Belicec, is obsessed with reading and saving the clippings about science and strange phenomena taken from major newspapers.

CHAPTER 5

1. *The Truth Machine* traces the life of Pete Armstrong, a prodigy whose childhood is tragically interrupted when a paroled killer murders his younger brother. Longing to correct the legal system that has failed his family, Pete de-

votes his life to the invention and deployment of a foolproof truth machine—one that could better determine potential or continued criminality. During the climax of the novel, just before the truth machine is to be approved by the U.S. government for use and distribution, Pete murders a colleague who has continually threatened his personal integrity, the timely completion of the truth machine, and Pete's ability to pass a final, governmental screening test. Because he invented the truth machine's software and maintains a monopoly on the patent and the market for twenty-five years, Armstrong is able to reprogram the instrument, allowing him to bypass its lie detection program by repeating "O Captain My Captain" whenever he is questioned. Unfortunately for Pete, after twenty-five years other truth machines come on the market; indeed, they are ubiquitous in society and are required for licenses of all kinds—including parenting. Armstrong is ultimately forced to decide between his own life and the integrity of the national surveillance system he has created. Because he cannot override the new versions of his truth machine, and because he does not want to endanger his family's integrity, Pete is forced to turn himself in—revealing the homicide he committed and covered up years before.

2. See also Wolfe et al. (2002) for a discussion of the "public acceptability of controversial technologies (PACT)," an STS analytic frame that could also be applied to brain-based technologies.

3. As a side note, I would cite an even older mention of the "organ of deceit" in *Outlines of Lectures on the Neurological System of Anthropology as Discovered, Demonstrated, and Taught in 1841 and 1842* by Joseph Rodes Buchanan; his is a phrenological reference that is tangentially related given its locationalist impulse (see, particularly, 332).

4. The reintroduction of biology as a legitimate and necessary term has largely been the work of feminist new materialists. For further information about feminist new materialism, see Sheridan (2002) and Hird (2003).

5. See the work of Mohamed et al. (2006), Langleben et al. (2002, 2005), Davatzikos et al. (2005), Spence et al. (2001, 2004), Kozel et al. (2004), Ganis et al. (2003), and Lee et al. (2002). See also O'Brien (2001).

6. The GKT was introduced by David Lykken in 1959. As the name suggests, it is not a lie detection test so much as a knowledge test; however, it has been used for interrogation and is often a central component to polygraphy tests.

7. A precursor to the GKT is the Control Question Test (CQT), which was developed by Leonarde Keeler and perfected as a method by John Reid in the late 1940s. See D. T. Lykken, "The GSR in the Detection of Guilt" (1959), and D. T. Lykken, "The Validity of the Guilty Knowledge Technique: The Effects of Faking" (1960).

8. For historical literature on the flaws of the polygraph, see Inbau (1957); for more contemporary critiques, see especially Lykken (1998) and National Research Council (2003).

9. The Society for Neuroscience *Brain Waves* publication is no longer available online. However, Langleben's statement is reiterated in a press release from his university (Penn). Therein, he is quoted explaining that "the results [of his 2001 experiment] indicate that since fMRI is a more direct measure of brain ac-

tivity than the method currently used in lie detection (the polygraph) it may have advantages over this technique" (O'Brien 2001).

10. Literature on the American security state post–9/11 tends to characterize new governmental surveillance as a trade-off between security and freedom (Paul 2003).

11. Lawrence Farwell and Daniel Langleben's brain-based detection techniques have proven to be lucrative market opportunities, particularly in a post–9/11 world. Economics bear this out: after certain risks became visible, the United States began "to increase its funding for research into novel kinds of lie detectors—raising the visibility of what had been an obscure corner of science" (Shulman 2002). Although agencies like the CIA, Department of Defense, Secret Service, and FBI "do not foresee using the Brain Fingerprinting technique for their operations," their hesitancy is based largely on the fact that Brain Fingerprinting has limited applications: it is "not designed as a screening tool" (Government Accountability Office 2001, 2). Despite the technology's narrow applicability, Farwell has still been quite successful in his sales pitch: he was given a two million dollar research grant from the CIA—which was subsequently withdrawn because he would not reveal his methods to agents—and anticipates that he could sell individual Brain Fingerprinting systems to law enforcement agencies for at least $100,000 each. He believes "the current atmosphere would be conducive to venture capital funding" (Cavuoto 2003) and admits, "Yes, I certainly hope I'll make money from this invention" (Brain Fingerprinting Laboratories 2001e). Private investors are already lining up, including James Halperin. The writer, and now patron of the sciences, invested in Brain Fingerprinting Laboratories in early 2004 (Solovitch 2004).

12. Importantly, this characterization carries over almost directly in the popular press as "the organ that's actually doing the lying" (Shulman 2002).

13. Anne Beaulieu and Joseph Dumit have addressed this shift toward a biological basis of mind in terms of trends in psychology and the scientific/popular interpretations of positron-emission tomography (PET) scans.

14. The biological mind is used heuristically, as the brain and mind are themselves both material objects and scientific constructs, and the relationship between neuroscience and psychology has always been a struggle as to who has the best research designs for access to information, the best working objects, and units of matter. I discuss one such debate in "Matter for Thought: The Psychon in Neurology, Psychology, and American Culture, 1927–1943" (2010).

15. Perhaps because of its problematic assumptions and implications, psychophysiology tends to be a negatively charged descriptor, rarely used by fMRI or EEG researchers. When the term is used, the implicit connections being drawn between biology and behavior are often ignored. Farwell's language is particularly illustrative. While he admits that Brain Fingerprinting is based in "cognitive psychophysiology," he defines this science as "the measurement of information processing activity in the brain" (Brain Fingerprinting Laboratories 2001d). Here he extrapolates "information processing" from electrical activity without reference to the underlying physiological measures of EEG.

16. For a more detailed explanation see D'Esposito, Deouell, and Gazzaley (2003).

17. The literal and metaphoric brain mapping used in fMRI is not a new phenomenon. Beginning with locationalist theory, introduced by Broca and Wernicke in the 1880s, areas of the brain involved in specific aspects of cognition like language and memory can be located as isolated areas of brain activity (Flourens 1824; Lashley 1937; Luria 1973). Standards for brain imaging emerged with the 1988 *Talaraich Atlas* and have been more recently modified by the MNI Brain (Cacioppo, Tassinary, and Berntson 2000; Brett 2005). This latest phase of brain mapping has meant a shift from fleshy brain banks to digital data banks (Beaulieu 2004). Norms established by brain data banks are problematic for many reasons: they blur distinctions between data collection and experimentation, they are being put to uses for which they were not originally intended, they have "normative potential" (Beaulieu 2004, 383), yet these data banks are not necessarily representative of any universally normal brain. The 1988 atlas is based on one woman's brain. The MNI brain is, at least, a composite of 305 brains, but even this amalgamation is biased: all of the participants were right-handed and between the ages of nineteen and twenty-six, and over 78 percent of the brains (239/305) were male.

18. One caveat: *when* you purchase a map does often matter; the date of the brain map may determine the level of detail, knowledge, and/or theories about the brain.

19. Catherine Waldby makes a similar point about anatomy in *The Visible Human Project* (2000): "The anatomical image can never be simply a benign reflection of the human body as referent, an illustration of its preexisting anatomical order" (55). While brain imaging techniques no longer require invasive procedures, the maps they use are based on dissection, destruction, and the reordering of biological materials (Sententia 2001; Beaulieu 2004).

20. Although a bit off topic for this argument, it should be noted that the sexual innuendo of Hanson's description is particularly relevant to and prevalent in histories of lie detection. William Marston performed theatrical experiments to determine the amorous nature of women (Marston 1917; "Blondes" 1928), and polygraphs have more recently been used to test for sexual deviance (Madsen, Parsons, and Grubin 2004; Wilcox and Sosnowski 2005).

21. In a later article (2005), Langleben et al. provide this reference: Augustine St (1948): "De mendacio." In *Opuscules. II. Problèmes moraux*, 244–45. Paris: Desclée de Brouwer.

22. See Thomas (1999) for more specifics on the collusion of foreignness and criminality—particularly in several early Sherlock Holmes adventures.

23. Farwell and his mentor, Emanuel Donchin, have published articles on the P300 in peer-reviewed journals (Farwell and Donchin 1988, 1991), and yet, Farwell has left academic publishing behind in favor of self-published essays available on his Web site. While Brain Fingerprinting's P300 wave has been well studied within the psychological community, has been related to alcoholism, and even has been used by Farwell as an aid for disabled people, its direct link to

guilty memories has not yet been proven. Responses to certain stimuli do not necessarily indicate criminal guilt. If, for example, a subject is shown a blue automobile, and his brain reacts, this does not necessarily mean he was the driver of the getaway car; he may simply have owned a blue car similar to the one pictured on the screen. Here, intentionality becomes a crucial factor in the detection test. Moreover, psychologists—including Emanuel Donchin—note the uncertainty of recall. Instead of the static, storage model employed by Farwell, Donchin argues that "memory is an active, creative process and not a passive repository of stored images" (Government Accountability Office 2001, 15). Theorists such as Elizabeth Wilson have addressed the problem of memory by arguing that "as information is copied and transferred through different memory states, the cognitive trace (or the mark of processing) is continuously being remade" (117). See also Schacter (1999) for perspectives on memory from psychology and cognitive neuroscience. Despite these dilemmas, Farwell maintains his ability to catalog brain content.

CODA

1. The polygraph test has been infamously compared to being attached to an electric chair. See, for example, John Reid and Richard Arther's "Behavior Symptoms of Lie Detection Subjects," *Journal of Criminal Law, Criminology, and Police Science:* "both the guilty and the innocent alike often make some half-humorous comment when entering the examining room, e.g., 'Boy, the electric chair,' or, 'Now I'll know how the hot seat feels'" (1953, 107).

# Works Cited

Adrian, E. D., and B. H. C. Matthews. 1934. The interpretation of potential waves in the cortex. *Journal of Physiology* 81 (4): 440–71.

Alder, Ken. 2002. A social history of untruth: Lie detection and trust in twentieth-century America. *Representations* 80:1–33.

Alder, Ken. 2007. *The lie detectors: The history of an American obsession.* New York: Free Press.

Ambach, Wolfgang, Rudolf Stark, Martin Peper, and Dieter Vaitl. 2008. Separating deceptive and orienting components in a Concealed Information Test. *International Journal of Psychophysiology* 70 (2): 95–104.

Appelbaum, Paul S. 2007. The new lie detectors: Neuroscience, deception, and the courts. *Psychiatry Services* 58 (4): 460–62.

Armstrong, Tim. 1998. *Modernism, technology, and the body: A cultural study.* Cambridge: Cambridge University Press.

Ashley, Mike. 2004. *The Gernsback days.* Rockville: Wildside Press.

Atwood, Margaret. 1981. *True Stories.* New York: Oxford University Press.

Augustine, St. 1847. On lying. In *Seventeen short treatises of Augustine, S., Bishop of Hippo.* Translated by John Henry Parker, 382–425. London: Oxford.

Balmer, Edwin. 1909. *Waylaid by wireless: A suspicion, a warning, sporting proposition, and a transatlantic pursuit.* Boston: Small, Maynard.

Balmer, Edwin, and Philip Wylie. 1933. *When worlds collide.* New York: Frederick A. Stokes Company.

Balmer, Edwin, and Philip Wylie. 1934. *After worlds collide.* New York: Frederick A. Stokes Company.

Balmer, Edwin, and William MacHarg. 1909a. The chalchihuitl stone. *Hampton's Magazine* 23 (5): 675–90.

Balmer, Edwin, and William MacHarg. 1909b. The empty cartridges. *Hampton's Magazine* 23 (6): 837–50.

Balmer, Edwin, and William MacHarg. 1909c. The fast watch. *Hampton's Magazine* 22 (6): 784–96.

Balmer, Edwin, and William MacHarg. 1909d. The man higher up. *Hampton's Magazine* 23 (4): 470–83.

Balmer, Edwin, and William MacHarg. 1909e. The man in the room. *Hampton's Magazine* 22 (5): 605–17.

Balmer, Edwin, and William MacHarg. 1909f. The private bank puzzle. *Hampton's Magazine* 23 (2): 179–91.

Balmer, Edwin, and William MacHarg. 1909g. The red dress. *Hampton's Magazine* 23 (1): 15–27.

Balmer, Edwin, and William MacHarg. 1910a. *The achievements of Luther Trant*. Boston: Small and Maynard.

Balmer, Edwin, and William MacHarg. 1910b. The Axton letters. *Hampton's Magazine* 24 (1): 94–107.

Balmer, Edwin, and William MacHarg. 1910c. The eleventh hour. *Hampton's Magazine* 24 (2): 243–55.

Balmer, Edwin, and William MacHarg. 1910d. The hammering man. *Hampton's Magazine* 24 (5): 705–16.

Balmer, Edwin, and William MacHarg. 1910e. A matter of mind reading. *Hampton's Magazine* 25 (4): 477–88.

Balmer, Edwin, and William MacHarg. 1915. Decidedly odd. *Top-Notch*. May 1, 75–88.

Balmer, Edwin, and William MacHarg. 1915. The day and the hour. *Top-Notch*. July 15, 166–81.

Balmer, Edwin, and William MacHarg. 1926. The man higher up. *Amazing Stories*. December.

Balmer, Edwin, and William MacHarg. 1927a. The eleventh hour. *Amazing Stories*. February.

Balmer, Edwin, and William MacHarg. 1927b. The hammering man. *Amazing Stories*. March.

Balmer, Edwin, and William MacHarg. 1927c. The man in the room. *Amazing Stories*. April.

Balmer, Edwin, and William MacHarg. 1930a. The eleventh hour. *Scientific Detective Monthly*. May.

Balmer, Edwin, and William MacHarg. 1930b. The fast watch. *Scientific Detective Monthly*. January.

Balmer, Edwin, and William MacHarg. 1930c. The hammering man. *Scientific Detective Monthly*. April.

Balmer, Edwin, and William MacHarg. 1930d. The man higher up. *Scientific Detective Monthly*. February.

Balmer, Edwin, and William MacHarg. 1930e. The man in the room. *Scientific Detective Monthly*. March.

Balmer, Edwin, and William MacHarg. 1915. Decidedly odd. *Top-Notch*. May 1, 75–88.

Balmer, Edwin, and William MacHarg. 1915. The day and the hour. *Top-Notch*. July 15, 166–81.

Barad, Karen. 2003. Posthumanist performativity: Toward an understanding of how matter comes to matter. *Signs* 28 (3): 801–31.

Beaulieu, Anne. 2000. The space inside the skull: Digital representations, brain mapping and cognitive neuroscience in the decade of the brain. PhD diss., McGill University, Montreal.

Beaulieu, Anne. 2002. Images are not the (only) truth: Brain mapping, visual knowledge, and iconoclasm. *Science, Technology, and Human Values* 27 (1): 53–86.

Beaulieu, Anne. 2003. Brains, maps, and the new territory of psychology. *Theory and Psychology* 13 (4): 561–68.

Beaulieu, Anne. 2004. From brainbank to database: The informational turn in the study of the brain. *Studies in History and Philosophy of Biological and Biomedical Science* 35:367–90.

Beck, Ulrich. 1992. *Risk society: Towards a new modernity.* London: Sage Publications.

Benussi, Vittorio. 1914. Die atmungssymptome der lüge [The respiratory symptoms of lying]. *Archiv für die gesamte psychologie* 32:50–57. Reprinted in *Polygrapyh* 4 (1): 52–76.

Bester, Alfred. 1996. *The demolished man.* New York: Vintage. Original edition, 1953.

Bisch, Louis. 1922. *Your inner self.* Garden City: Doubleday, Page.

Blondes lose out in fiim love test. 1928. *New York Times,* January 31, 25.

Blumenthal, Jeremy. 2002. Law and social science in the twenty-first century. *Southern California Interdisciplinary Law Journal* 12 (1): 1–53.

*Body Snatchers.* DVD. 1994. Directed by Abel Ferrara. Warner Bros.

Boire, Richard. 2005. Searching the brain: The fourth amendment implications of brain-based deception detection. *American Journal of Bioethics* 5 (2): 62–63.

Bok, Sissela. 1978. *Lying: Moral choice in public and private life.* New York: Pantheon. Vintage, updated edition, 1999.

Borck, Cornelius. 2001. Electricity as a medium of psychic life: Electrotechnological adventures into psychodiagnosis in Weimar Germany. *Science in Context* 14 (4): 565–90.

Bourne, Frank. 1930. The thought stealer. *Science Fiction Series* 7:3–13.

Bowden, Mark. 2003. The dark art of interrogation. *Atlantic Monthly,* October, 51–76.

Brain, Robert. 1996. The graphic method: Inscription, visualization, and measurement in nineteenth-century science and culture. PhD diss., University of California.

Brain, Robert. 2002. Representations on the line: Graphic recording instruments and scientific modernism. In *From energy to information: Representation in science and technology, art, and literature,* ed. Bruce Clarke and Linda Henderson. Stanford: Stanford University Press.

Brain Fingerprinting Laboratories, Inc. 2001a. Brain fingerprinting on the Discovery Channel: The new detectives; Case studies in forensic science. http://www.brainwavescience.com/discovery.php.

Brain Fingerprinting Laboratories, Inc. 2001b. Brain Fingerprinting testing ruled admissible in court. http://www.brainwavescience.com/Ruled%20Admissable.php.

Brain Fingerprinting Laboratories, Inc. 2001c. *Counterterrorism applications.* http://www.brainwavescience.com/counterterrorism.php.

Brain Fingerprinting Laboratories, Inc. 2001d. *Highlights of CBS 60 Minutes featuring Brain Fingerprinting.* http://www.brainwavescience.com/CBS.

Brain Fingerprinting Laboratories, Inc. 2001e. Interview with Dr. Lawrence Farwell: Frequently asked questions about brain fingerprinting testing. http://www.brainwavescience.com/FreqAskedQuestions.php.

Brain Fingerprinting Laboratories, Inc. 2001f. Questionnaire. http://www.brainwavescience.com/Questionnaire.htm. Accessed May 16, 2007.

Brennan, Teresa. 2004. *The transmission of affect.* Ithaca: Cornell University Press.

Brett, Matthew. 2005. *The MNI Brain and the Talairach Atlas.* Medical Research Council homepage: cognition and brain sciences unit imagine homepage. http://www.mrc-cbu.cam.ac.uk/Imaging/Common/mnispace.shtml. Accessed January 27, 2007.

Breuer, William. 1995. *J. Edgar Hoover and his G-men.* Westport: Praeger.

Browne, Douglas, and Alan Brock. 1953. *Fingerprints: Fifty years of scientific crime detection.* London: George G. Harrap.

Brumberg, Joan Jacobs. 1997. *Body project: An intimate history of American girls.* New York: Random House.

Buchanan, Joseph Rhodes. 1854. *Outlines of lectures on the neurological system of anthropology as discovered, demonstrated, and taught in 1841 and 1842.* Cincinnati: The Office of Buchanan's Journal of Man.

Buller, Tom. 2005. Can we scan for truth in a society of liars? *American Journal of Bioethics* 5 (2): 58–60.

Bunn, Geoffrey. 1997. The lie detector, *Wonder Woman,* and liberty: The life and work of William Moulton Marston. *History of the Human Sciences* 10 (1): 91–119.

Bunn, Geoffrey. 1998. The hazards of the will to truth: A history of the lie detector. PhD diss., York University.

Bunn, Geoffrey. 2007. Spectacular science: The lie detector's ambivalent powers. *History of Psychology* 10 (2): 156–78.

Burnham, John. 1988. *Paths into American culture: Psychology, medicine, and morals.* Philadelphia: Temple University Press.

Cacioppo, John, Louis Tassinary, and Gary Berntson, eds. 2000. *Handbook of psychophysiology.* 2nd ed. Cambridge: Cambridge University Press.

Cairns, Huntington. 1935. *Law and the social sciences.* New York: Harcourt, Brace, and Howe.

Callon, Michael, and Bruno Latour. 1981. Unscrewing the big Leviathan: How actors macrostructure reality and how sociologists help them to do so. In *Advances in social theory and methodology: Toward an integration of micro- and macrosociologies,* ed. K. Knorr-Cetina and A. V. Cicourel, 277–303. Boston: Routledge and Kegan Paul.

Camp stories contest: The lie detector. 1918. *Chicago Daily Tribune,* February 17, B6.

Campbell, Jeremy. 2001. *The liar's tale: A history of falsehood.* New York: W. W. Norton.

Campbell, John. 1930. Solarite. *Amazing Stories.* November.

Campbell, Nancy. 2005. Suspect technologies: Scrutinizing the intersection of science, technology, and policy. *Science, Technology, and Human Values* 30:374–402.

Cannon, Walter. 1915. *Bodily changes in pain, hunger, fear, and rage.* New York: D. Appleton.

Carte, Gene, and Elaine Carte. 1975. *Police reform in the United States: The era of August Vollmer, 1905–1932.* Berkeley: University of California Press.

Cartwright, Lisa. 1992. Experiments of destruction: Cinematic inscriptions of physiology. Special issue: Seeing Science. *Representations* 40:129–52.

Cavuoto, James. 2003. *Brainwave sensor touted as tool in counter-terrorism.* Neurotech Business Report. www.neurotechreports.com/pages/brainfinger printing.html.

Cheng, Edward K., and Albert H. Yoon. 2005. Does Frye or Daubert matter? A study of scientific admissibility standards. *Virginia Law Review* 90:471–513.

Clarke, Bruce. 2001. *Energy forms: Allegory and science in the era of classical thermodynamics.* Ann Arbor: University of Michigan Press.

Clarke, Bruce, and Linda Henderson. 2002. *From energy to information: Representation in science and technology, art, and literature.* Stanford: Stanford University Press.

Clue to chemistry of heredity found. 1953. *New York Times,* June 13, 17.

Cohen, Benjamin. 2004. The elements of the table: Visual discourse and the periodic representation of chemical classification. *Configurations* 12 (1): 41–76.

Cole, Simon. 2001. *Suspect identities: A history of fingerprinting and criminal identification.* Cambridge, MA: Cambridge University Press.

Coleman, Linda, and Paul Kay. 1981. Prototype semantics: The English word lie. *Language* 57 (1): 26–44.

Conan, Neal. 2006. The future of lie detecting. In *Talk of the Nation,* National Public Radio, June 26.

Cooke, T. G. 1941. *The blue book of crime.* Chicago: Institute of Applied Science.

Crary, Jonathan. 1990. *Techniques of the Observer: On vision and modernity in the 19th century.* Cambridge: MIT Press.

Crary, Jonathan. 2000. *Suspensions of perception: Attention, spectacle, and modern culture.* Cambridge: MIT Press.

Cutmore, Tim, Tatjuana Djakovic, Mark Kebbell, and David Shum. 2008. An object cue is more effective than a word in ERP-based detection of deception. *International Journal of Psychophysiology* 71 (3): 185–92.

Damasio, Antonio. 1994. *Descartes's error: Emotion, reason, and the human brain.* New York: Penguin.

Daston, Lorraine, and Peter Galison. 1992. The Image of Objectivity. Special issue: Seeing science, *Representations* 40:81–128.

Daston, Lorraine, and Peter Galison. 2007. *Objectivity.* New York: Zone.

Davatzikos, Christos, Kosha Ruparel, Yong Fan, Dinggang Shen, M. Acharyya, James Loughead, Ruben Gur, and Daniel Langleben. 2005. Classifying spatial patterns of brain activity with machine learning methods: Application to lie detection. *Neuroimage* 28 (3): 663–68.

D'Esposito, Mark, Leon Y. Deouell, and Adam Gazzaley. 2003. Alterations in the

BOLD FMRI signal with ageing and disease: A challenge for neuroimaging. *Nature Reviews: Neuroscience* 4:863–72.

Dick, Philip K. 1996. *Do androids dream of electric sheep?* New York: Del Ray. Original edition, 1968.

Dick, Philip K. 2002a. The imposter. In *Selected stories of Philip K. Dick.* New York: Pantheon. Original edition, 1953.

Dick, Philip K. 2002b. *The simulacra.* New York: Vintage. Original edition, 1964.

Dimock, Wai Chee, and Priscilla Wald. 2002. Literature and science: Cultural forms, conceptual exchanges. *American Literature* 74 (4): 705–14.

Display Ad 9, No Title. *Chicago Daily Tribune.* 1910. May 21, 12.

Dror, Otniel. 1998. Creating the emotional body: Confusion, possibilities, and knowledge. In *An emotional history of the United States,* ed. Peter Stearns and Jan Lewis. New York: New York University Press.

Dror, Otniel. 1999a. The affect of experiment: The turn to emotions in Anglo-American physiology, 1900–1940. *Isis* 90 (2): 205–37.

Dror, Otniel. 1999b. The scientific image of emotion: Experience and technologies of inscription. *Configurations* 7 (3): 355–401.

Dror, Otniel. 2001a. Counting the affects: Discoursing in numbers. *Social Research* 68 (2): 357–78.

Dror, Otniel. 2001b. Techniques of the brain and the paradox of emotions, 1880–1930. *Science in Context* 14 (4): 643–60.

Dumit, Joseph. 2003. *Picturing personhood: Brain scans and biomedical identity.* Information series. Princeton: Princeton University Press.

Dunbar, Robin. 1996. *Grooming, gossip, and the evolution of language.* Cambridge: Harvard University Press.

Ebertle, Merab. 1930. The thought translator. *Science Fiction Series* 9:3–18.

Ehrenwald, Jan. 1948. *Telepathy and medical psychology.* New York: W. W. Norton.

Ekman, Paul. 1985. *Telling lies: Clues to deceit in the marketplace, politics, and marriage.* New York: W. W. Norton.

Ellis, Havelock. 1913. *The criminal.* London: Walter Scott. Original edition, 1890.

Ernst, Paul. 1933. From the wells of the brain. *Astounding Stories* 35:48–54.

Farwell, L. A. 2000. Brain Fingerprinting: Brief summary of the technology. Brain Fingerprinting Laboratories, Inc. http://www.forensic-evidence.com/site/Behv_Evid/Farwell_sum6_oo.html.

Farwell, L. A. 2001a. Brain fingerprinting laboratories: Scientific procedure, research, and applications. Brain Fingerprinting Laboratories, Inc. http://www.brainwavescience.com/TechnologyOverview.php.

Farwell, L. A. 2001b. Brain fingerprinting testing and memory issues. Brain Fingerprinting Laboratories, Inc. http://www.brainwavescience.com/MemoryIssues.php.

Farwell, L. A. 2003. The scope and science of brain fingerprinting testing: Scientific data and its relationship to findings of fact and law. Brain Fingerprinting Laboratories, Inc. http://www.brainwavescience.com/ScopeandScienceofBF.php.

Farwell, L. A., and Emanuel Donchin. 1988. Event-related potentials in interrogative polygraphy: Analysis using bootstrapping. *Psychophysiology* 25 (4): 445.

Farwell, L. A., and Emanuel Donchin. 1991. The truth will out: Interrogative polygraphy ("lie detection") with event-related potentials. *Psychophysiology* 28 (5): 531–47.

Feder, Barnaby. 2001. Truth and justice, by the blip of a brainwave. *New York Times*. http://www.brainfingerprinting.com/NewYorkTimes.php.

Finney, Jack. 1996. *Invasion of the body snatchers*. New York: Simon and Schuster. Original edition, 1955.

Fisher, Jim. 1994. *The Lindbergh case*. Piscataway: Rutgers University Press.

Flourens, Pierre. 1824. *Recherches expérimentates sur les propriétés et les fonctins du système nerveux dans les animaux vertébrés*. Paris: Chez Crevot.

Ford, Elizabeth. 2006. Lie detection: Historical, neuropsychiatric, and legal dimensions. *International Journal of Law and Psychiatry* 29 (3): 159–77.

Foucault, Michel. 1995. *Discipline and punish: The birth of the prison*. Trans. Alan Sheridan. New York: Vintage.

Frank, Jerome. 1930. *Law and the modern mind*. New York: Brentano's.

Frank, Lone. 2007. *Mindfield: How brain science is changing our world*. Oxford: Oneworld Publications.

Franklin, Sarah, Celia Lury, and Jackie Stacey. 2000. *Global nature, global culture*. London: Sage.

Freeman, R. Austin. 1909. *John Thorndyke's cases*. London: Chatto and Windus.

Freud, Sigmund. 1929. *Civilization and its discontents*. New York: Dover.

Frith, Susan. 2004. Who's minding the brain? *Pennsylvania Gazette* (3). http://www.upenn.edu/gazette/0104/frith1.html.

*Frye v. United States*. 1923. 54 App. D.C. 46, 293 F. 1013.

Galison, Peter. 1997. *Image and logic: A material culture of microphysics*. Chicago: University of Chicago Press.

Galton, Francis. 1879. Psychometric experiments. *Brain* 2:149–62.

Galton, Francis. 1892. *Finger prints*. London: Macmillan.

Gammage, Jeff. 2002. 9/11 one year later: Gearing up. *Philadelphia Inquirer*. Sept. 7. http://www.philly.com/mld/inquirer/news/special_packages/sept11/4017725.htm. Accessed January 27, 2007.

Ganis, G., S. M. Kosslyn, S. Stose, W. L. Thompson, and D. A. Yurgelun-Todd. 2003. Neural correlates of different types of deception: An fMRI investigation. *Cerebral Cortex* 13:830–36. Accessed January 27, 2007.

Geddes, L. A. 1974. History of the polygraph, an instrument for the detection of deception. *Biomedical Engineering*, 154–56.

Geddes, L. A. 2002. The truth shall set you free. *IEEE* 21 (3): 97–100.

Gernsback, Hugo. 1926. A new sort of magazine. *Amazing Stories*. April.

Gernsback, Hugo. 1930a. How to write "science" stories. *Writer's Digest* 10:27–29.

Gernsback, Hugo. 1930b. Science vs. crime. *Scientific Detective Monthly* 1 (1): 84–85.

Gloor, Pierre. 1969. *Hans Berger and the electroencephalogram of man: The fourteen original reports on the human encephalogram*. Vol. 28, *Electroencephalography and Clinical Neurophysiology Supplement*. Amsterdam: Elsevier.

Gloor, Pierre. 1994. Berger lecture: Is Berger's dream coming true? *Electroencephalography and Clinical Neuroscience* 90:253–66.

Golan, Tal. 2004. *Laws of men and laws of nature: The history of scientific expert testimony in England and America.* Cambridge: Harvard University Press.

Goldberg, Steven. 1994. *Culture clash: Law and science in America.* New York: New York University Press.

Goldstein, Eva. 1923. Reaction times and the consciousness of deception. *American Journal of Psychology* 34 (4): 562–81.

Gould, Stephen J. 1996. *The mismeasure of man.* New York: W. W. Norton.

Government Accountability Office. 2001. Investigative techniques: Federal agency views on the potential application of "Brain Fingerprinting." www.gao.gov/new.items/d0222.pdf.

Greasley, Philip. 2001. *Dictionary of Midwestern literature.* Vol. 1, *The authors.* Bloomington: Indiana University Press.

Guisey, J. U. 1917. The unknown quantity. *All-Story Weekly* 75:290–307.

Guisey, J. U. 1919. The ivory pipe. *All-Story Weekly* 101:529–67.

Hale, Matthew. 1980. *Human science and social order: Hugo Münsterberg and the origins of applied psychology.* Philadelphia: Temple University Press.

Hall, Carl. 2001. Fib detector: Study shows brain scan detects patterns of neural activity when someone lies. *San Francisco Chronicle,* November 26, A-10.

Hall, T. Proctor. 1929. Dr. Glee's Experiments. *Amazing Stories Quarterly* 2 (3).

Halperin, James. 1996. *The truth machine.* New York: Ballantine.

Hamilton, Heather G. 1998. The movement from *Frye* to *Daubert:* Where do the states stand? *Jurimetrics* 98:210–13.

Hanson, Allen. 1992. *Testing testing: Social consequences of the examined life.* Berkeley: University of California Press.

Happell, Mark D. 2005. Neuroscience and the detection of deception. *Review of Policy Research* 22 (5): 667–85.

Haraway, Donna. 1991. *Simians, cyborgs, and women: The reinvention of nature.* London: Routledge.

Haraway, Donna. 1997. *Modest Witness@Second-Millennium.FemaleMan-Meets-OncoMouse.* New York: Routledge.

Harding, Sandra. 1991. *Whose science? Whose knowledge? Thinking from women's lives.* Ithaca: Cornell University Press.

Hayles, N. Katherine. 1999. *How we became posthuman: Virtual bodies in cybernetics, literature, and informatics.* Chicago: University of Chicago Press.

Heinlein, Robert. 1986. *The puppet masters.* New York: Del Ray. Original edition, 1951.

Henderson, Linda. 2002. Vibratory modernism: Boccioni, Kupka, and the ether of space. In *From energy to information: Representation in science and technology, art, and literature,* ed. Bruce Clarke and Linda Henderson. Stanford: Stanford University Press.

Henke, F. G., and M. W. Eddy. 1909. Mental diagnosis by the association reaction method. *Psychological Review* 16 (5): 399–409.

Henseler, T. B. 1997. Comment: A critical look at the admissibility of polygraph evidence in the wake of Daubert: The lie detector fails the test. *Catholic University Law Review* 46:1247–48.

Herschel, William. 1916. *The Origins of finger-printing.* London: Oxford University Press.

Herzig, Rebecca. 2004. On performance, productivity, and vocabularies of motive in recent studies of science. *Feminist Theory* 5 (2): 127–47.

Hesse, Mary. 1970. *Models and analogies in science.* Notre Dame: University of Notre Dame Press.

Hird, Mira. 2003. From the matter of culture to the culture of matter: Feminist explorations of nature and science. *Sociological Research Online* (1). http://www.socresonline.org.uk/8/1/hird.html.

Hoberman, J. 1994. Paranoia and the pods. *Sight and Sound* 4 (5): 28–31.

Holmes, Thomas. 1912. *Psychology and crime.* London: J. M. Dent and Sons.

Hoover, J. Edgar. 1935a. Modern problems of law enforcement. Presented at *The Convention of the International Association of Chiefs of Police.* July 29. Atlantic City, NJ.

Hoover, J. Edgar. 1935b. Scientific methods of crime detection in the judicial process. *George Washington Law Review* 4 (1): 1–25.

Hoover, J. Edgar. 1936a. The influence of crime on the American home. Address before the *Round Table forum under the auspices of the New York Herald Tribune* at New York City, March 11. Washington, U.S. Govt. Print Off.

Hoover, J. Edgar. 1936b. Local law enforcement in relation to national crime. Presented at *The Sheriffs and Peace Officers Association of Oklahoma.* Tulsa, OK.

Hoover, J. Edgar. 1936c. Patriotism and the war against crime. An address given before the annual convention of the *Daughters of the American Revolution.* Washington, DC, April 23, from the office of Congressional and Public Affairs (OCPA), FBI.

Hoover, J. Edgar. 1936d. Science in law enforcement. An address delivered before the *Annual Convention of the International Association for Identification,* September 29. Dallas, TX: Department of Justice.

Hoover, J. Edgar. 1937. Criminal identification and the functions of the identification division. Washington: Federal Bureau of Investigation.

Hoover, J. Edgar. 1938. Progress in crime control: Columbia Broadcasting System.

Hoover, J. Edgar. 1956. *Masters of deceit: The story of communism in America and how to fight it.* New York: Henry Holt.

Illes, Judy. 2003. Neuroethics in a new era of neuroimaging. *American Journal of Neuroradiology* 24 (9): 1734–41.

Inbau, Fred. 1934. Scientific evidence in criminal cases: Methods of detecting deception. *Journal of Criminal Law and Criminology* 24 (6): 1140–58.

Inbau, Fred. 1935a. The admissibility of scientific evidence in criminal cases. *Law and Contemporary Problems* 2:495–503.

Inbau, Fred. 1935b. Detection of deception technique admitted as evidence. *Journal of the American Institute of Criminal Law and Criminology* 26:262–70.

Inbau, Fred. 1935c. The lie detector. *Scientific Monthly* 46 (1): 81–87.

Inbau, Fred. 1938. Book review: *The lie detector test* by William Marston. *Journal of the American Institute of Criminal Law and Criminology* 29:305–7.

Inbau, Fred. 1942. *Lie detection and criminal interrogation*. Baltimore: Waverly.

Inbau, Fred. 1957. Lie-detector test limitations. *Journal of Forensic Sciences* 2 (3): 255–62.

Inman, Keith, and Norah Rudin. 2000. *Principles and practices of criminalistics: The profession of forensic science*. Boca Raton: CRC Press.

*The Invasion*. DVD. 2007. Directed by Oliver Hirschbiegel. Warner Bros.

*Invasion of the Body Snatchers*. DVD. 1954. Directed by Don Siegel. Walter Wagner Production.

*Invasion of the Body Snatchers*. DVD. 1978. Directed by Phillip Kaufman. United Artists.

Johnson, Glen M. 1979. "We'd fight . . . we had to": *The Body Snatchers* as novel and film. *Journal of Popular Culture* 13:5–16.

Johnson, Steve. 2002. *Emergence: The connected lives of ants, brains, cities, and software*. New York: Simon and Schuster.

Joyce, Kelly A. 2008. *Magnetic appeal: MRI and the myth of transparency*. Ithaca: Cornell University Press.

Jung, Carl. 1906. The association method. *American Journal of Psychology* 31:219–69.

Kauffman, H. H. 2001. The expert witness: Neither *Frye* nor *Daubert* solved the problem: What can be done? *Science and Justice* 41 (1): 7–20.

Keeler, Leonarde. 1930. A method for detecting deception. *American Journal of Police Science* 1:38–51.

Keeler, Leonarde. 1934. Debunking the "lie detector." *Journal of Criminal Law and Criminology* 25 (1): 153–59.

Keeley, James P. 2001. Subliminal promptings: Psychoanalytic theory and the Society for Psychical Research. *American Imago* 58 (4): 767–91.

Keller, Evelyn Fox. 1992. *Secrets of life, secrets of death: Essays on language, gender, and science*. New York: Routledge.

Keller, Evelyn Fox, and Helen Longino. 1986. *Feminism and science*. London: Falmer Press.

Kelly, Forence. 1935. Science and the war on crime. *New York Times*, September 15, BR4.

Kennedy, Donald. 2005. Neuroimagine: Revolutionary research tool or a postmodern phrenology. *American Journal of Bioethics* 5 (2): 19.

Kirsch, Steve. 2001. *Identifying terrorists before they strike by using computerized knowledge assessment (CKA)*. Steve Kirsch's Political Home Page 2001. http://www.skirsch.com/politics/plane/ultimate.htm.

Kozel, Andrew, Letty Revell, Jeffery Lorberbaum, Ananda Shastri, Jon Elhai, Michael Horner, Adam Smith, Ziad Nahas, Daryl Bohning, and Mark George. 2004. A pilot study of functional magnetic resonance imaging brain correlates of deception in healthy young men. *Journal of Neuropsychiatry and Clinical Neuroscience* 16 (3): 295–305.

Krasner, Leonard. 1983. The psychology of mystery. *American Psychologist* 38 (5): 578–82.

Kuhn, Thomas. 1962. *The structure of scientific revolutions*. Chicago: University of Chicago Press.

Kulynych, Jennifer. 2002. Legal and ethical issues in neuroimaging research: Human subjects protection, medical privacy, and the public communication of research results. *Brain and Cognition* 50 (3): 345–57.

Landy, Frank. 1992. Hugo Münsterberg: Victim or visionary? *Journal of Applied Psychology* 77 (6): 787–802.

Langleben, Daniel, James Loughead, Warren Bilker, Kosha Ruparel, Anna Childress, Samantha Busch, and Ruben Gur. 2005. Telling truth from lie in individual subjects with fast event-related fMRI. *Human Brain Mapping* 26 (4): 262–72.

Langleben, Daniel, L. Schroeder, J. A. Maldjian, R. C. Gur, S. McDonald, J. D. Ragland, C. P. O'Brien, and A. R. Childress. 2002. Brain activity during simulated deception: An event-related functional magnetic resonance study. *NeuroImage* 15 (3): 727–32.

Larson, John. 1921. Modification of the Marston deception test. *Journal of the American Institute of Criminal Law and Criminology* 12 (3): 390–99.

Larson, John. 1932. *Lying and its detection: A study of deception and deception tests.* Chicago: University of Chicago Press.

Lashley, Karl Spencer. 1937. Functional determinants of cerebral localization. *Arch Neural Psychiatry* 38 (2): 371–87.

Lassiter, G. Daniel. 2004. *Interrogations, confessions, and entrapment.* New York: Springer.

Latour, Bruno. 1987. *Science in action: How to follow scientists and engineers through society.* Cambridge: Harvard University Press.

Latour, Bruno. 1993. *We have never been modern.* Trans. Catherine Porter. Cambridge: Harvard University Press.

Latour, Bruno, and Steve Woolgar. 1979. *Laboratory life: The construction of scientific facts.* Princeton: Princeton University Press.

Laurence, William. 1935. Electricity in the brain records a picture of the action of thoughts. Special to the *New York Times,* April 14, 32.

Law, John, and Vicky Singleton. 2000. Performing technology's stories: On social constructivism, performance, and performativity. *Technology and Culture* 41 (4): 765–75.

Lawrence, William. 1954. Science analyzes heredity factors. *New York Times,* April 27, 31.

Lee, Henry, and R. E. Gaensslen. 1991. *Advances in fingerprint technology.* New York: Elsevier Science.

Lee, Tatiana, Ho-Ling Liu, Li-Hai Tan, Chetwyn Chan, Srikanth Mahankali, Ching-Mei Feng, Jinwen Hou, Peter T. Fox, and Jia-Hong Gao. 2002. Lie detection by functional magnetic resonance imaging. *Human Brain Mapping* 15 (3): 727–32.

Leo, Richard A. 1992. From coercion to deception: The changing nature of police interrogation in America. *Crime, Law, and Social Change* 18 (1–2): 35–59.

LeVay, Simon. 1994. *The sexual brain.* Cambridge: MIT Press.

Littlefield, Melissa. 2009. Constructing the organ of deceit: The rhetoric of fMRI and Brain Fingerprinting in post–9/11 America. *Science, Technology, and Human Values* 34:365–92.

Littlefield, Melissa. 2010. Matter for thought: The psychon in neurology, psychology, and American culture, 1927–1943. In *Neurology and modernity: A cultural history of nervous systems, 1800–1950,* ed. Laura Salisbury and Andrew Shail, 267–86. Houndmills: Palgrave.

Locard, Edmund. 1930. The analysis of dust traces Part I–III. *American Journal of Police Science* 1:276, 401, 496. Reprint.

Lombroso, Cesare. 1911. *Criminal man.* New York: C. P. Putnam's Sons. Original edition, 1887.

Luckhurst, Roger. 2002. *The invention of telepathy, 1870–1901.* New York: Oxford University Press.

Lui, Ming, and Peter Rosenfeld. 2008. Detection of deception about multiple, concealed, mock crime items, based on a spatial-temporal analysis of ERP amplitude and scalp distribution. *Psychophysiology* 45 (5): 721–30.

Lui, Ming Ann, J. Peter Rosenfeld, and Andrew H. Ryan Jr. 2008. Thirty-site P300 scalp distribution, amplitude variance across sites, and amplitude in detection of deceptive concealment of multiple guilty items. *Social Neuroscience* 28:1–19.

Luria, Alexander. 1973. *The working brain.* New York: Basic Books.

Lykken, David. 1959. The GSR in the detection of guilt. *Journal of Applied Psychology* 43:385–88.

Lykken, David. 1960. The validity of the guilty knowledge technique: The effects of faking. *Journal of Applied Psychology* 44:258–62.

Lykken, David. 1998. *A tremor in the blood: The uses and abuses of the lie detector.* New York: Plenum Press.

Lyon, Janet. 1999. *Manifestoes: Provocations of the modern.* Ithaca: Cornell University Press.

MacHarg, William. 1898. A Christmas fantasy. *Chicago Daily Tribune,* December 18, 60.

MacHarg, William. 1899. Mr. Dudd of Chicago. *Chicago Daily Tribune,* June 25, 45.

MacHarg, William. 1940. *The affairs of O'Malley.* New York: Dial Press.

Madsen, William, Shaun Parsons, and Don Grubin. 2004. A preliminary study of the contribution of periodic polygraph testing to the treatment and supervision of sex offenders. *Journal of Forensic Psychiatry and Psychology* 15 (4): 682–95.

Mann, Katrina. 2004. "You're next!": Postwar hegemony besieged in *Invasion of the Body Snatchers. Cinema Journal* 44 (1): 49–68.

Marston, William. 1917. Systolic blood pressure symptoms of deception. *Journal of Experimental Psychology* 2 (2): 117–63.

Marston, William. 1920. Reaction-time symptoms of deception. *Journal of Experimental Psychology* 3 (1): 72–87.

Marston, William. 1921a. Psychological possibilities in the deception tests. *Journal of the American Institute of Criminal Law and Criminology* 11 (4): 551–70.

Marston, William. 1921b. Systolic blood pressure and reaction time symptoms of deception and constituent mental states. PhD diss., Harvard University.

Marston, William. 1923. Sex characteristics of systolic blood pressure behavior. *Journal of Experimental Psychology* 6 (6): 387–419.

Marston, William. 1924. A theory of emotions and affection based upon systolic blood pressure studies. *American Journal of Psychology* 35 (4): 469–506.

Marston, William. 1924–25. Studies in testimony. *American Institute of Criminal Law and Criminology* 15:5–31.

Marston, William. 1925. Negative type reaction-time symptoms of deception. *Psychological Review* 32 (3): 241–47.

Marston, William. 1927. Motor consciousness as a basis for emotion. *Journal of Abnormal and Social Psychology* 22:140–50.

Marston, William. 1928. *Emotions of normal people.* London: Kegan Paul, Trench, Trubner.

Marston, William. 1929. Bodily symptoms of elementary emotions. *Psyche* 10:70–86.

Marston, William. 1938a. *The lie detector test.* New York: Richard R. Smith.

Marston, William. 1938b. New facts about shaving revealed by lie detector! *Time* 17:29.

Marston, William. 1943. Why 100,000,000 Americans read comics. *American Scholar* (Winter): 42–43.

Masden, L., S. Parsons, and D. Grubin. 2004. A preliminary study of the contribution of periodic polygraph testing to the treatment and supervision of sex offenders. *Journal of Forensic Psychiatry and Psychology* 15:682–95.

Masip, Jaume, Eugenio Garrido, and Carmen Herrero. 2004. Defining deception. *Anales de Psicología* 20 (1): 147–71.

McCall, Becky. 2004. Brain fingerprints under scrutiny. *BBC News.* http://news.bbc.co.uk/1/hi/sci/tech/3495433.stm.

McCarty, D. G. 1929. *Psychology for the lawyer.* New York: Prentice-Hall.

Millett, David. 2001. Hans Berger: From psychic energy to EEG. *Perspectives in Biology and Medicine* 44 (4): 522–42.

Moffett, Cleveland. 1909. *Through the Wall.* New York: Grosset and Dunlap.

Mohamed, Feroze, Scott Faro, Nathan Gordon, Steven Platek, Harris Ahmad, and J. Michael Williams. 2006. Brain mapping of deception and truth telling about an ecologically valid situation: Functional MR imaging and polygraph investigation—initial experience. *Radiology* 238 (2): 679–88.

Moore, Charles. 1907. Yellow psychology. *Law Notes* 11:125–27.

Morrison, Mark. 2001. *The public face of modernism: Little magazines, audiences, and reception, 1905–1920.* Madison: University of Wisconsin Press.

Münsterberg, Hugo. 1893. *Psychological laboratory of Harvard University.* Cambridge: University of Cambridge, MA.

Münsterberg, Hugo. 1908. *On the witness stand: Essays on psychology and crime.* New York: Doubleday, Page. Reprinted 1925.

National Research Council. 2003. *The polygraph and lie detection.* Committee to review the scientific evidence of the polygraph. Division of Behavioral and Social Sciences and Education. Washington, D.C.: The National Academies Press.

Nickell, Joe, and John F. Fischer. 1999. *Crime science: Methods of forensic detection.* Lexington: University of Kentucky Press.

Obermann, C. E. 1939. The effect on the Berger rhythm of mild affective states. *Journal of Abnormal and Social Psychology* 34 (1): 84–95.

O'Brien, Ellen. 2001. *The brain operates differently in deception and honesty.* University of Pennsylvania press release, http://www.eurekalert.org/pub_re leases/2001-11/uopm-tbo110901.php. November 11.

Obuchowski, Mary DeJung. 1995. *The Indian Drum* and its authors: A reconsideration. *The Yearbook of the Society for the Study of Midwestern Literature,* ed. David Anderson, 60–68. East Lansing: Midwestern Press.

Olson, Steve. 2005. Neuroimaging: Brain scans raise privacy concerns. *Science* 307 (5715): 1548–50.

Oppenheim, James. 1923. *Your hidden powers.* New York: Alfred A. Knopf.

Otis, Laura. 2001. *Networking: Communicating with bodies and machines in the nineteenth century.* Ann Arbor: University of Michigan Press.

Paul, Ron. 2003. Trading freedom for security: Drifting toward a police state. *Mediterranean Quarterly* 14 (1): 6–24.

Potter, Claire Bond. 1998. *War on crime: Bandits, G-men, and the politics of mass culture.* Brunswick: Rutgers University Press.

Powers, Richard. 1983. *G-men: Hoover's F.B.I. in American popular culture.* Carbondale: Southern Illinois University Press.

Prasad, Amit. 2005. Making images/making bodies: Visibilizing and disciplining through magnetic resonance imaging (MRI). *Science, Technology, and Human Values* 30 (2): 291–316.

Proctor, Robert. 1996. *The cancer wars: How politics shapes what we know and don't know about science.* New York: Basic Books.

Proctor, Robert, and Londa Schiebinger. 2007. *Agnotology: The making and unmaking of ignorance.* Palo Alto: Stanford University Press.

Rabinbach, Anson. 1990. *The human motor: Energy, fatigue, and the rise of modernity.* New York: Basic Books.

Rabinow, Paul. 1992. Galton's regret: Of types and individuals. In *DNA on trial: Genetic identification and criminal justice,* ed. Paul R. Billings, 5–18. New York: Cold Spring Laboratory Press.

Reeve, Arthur. 1912. *The silent bullet: The adventures of Craig Kennedy.* New York: Dodd, Mead.

Reid, John. 1947. A revised questioning technique in lie-detection tests. *Journal of Criminal Law and Criminology* 37 (6): 542–47.

Reid, John, and Richard Arther. 1953. Behavior symptoms of lie-detection subjects. *Journal of Criminal Law, Criminology, and Police Science* 44 (1): 104–8.

Reilly, John, ed. 1985. *Twentieth century crime and mystery writers.* New York: Palgrave.

Rhine, J. B. 1948. *The reach of the mind.* New York: William Sloane Associates.

Rhodes, Molly. 2000. Wonder Woman and her disciplinary powers. In *Doing science + culture,* ed. Roddey Reid. New York: Routledge.

Robinson, Edward. 1935. *Law and the lawyer.* New York: Macmillan.

Robinson, Henry Morton. 1935. *Science catches the criminal.* New York: Blue Ribbon Books.

Robinson, Jane Howard, Gene Carte, and Elaine Carte. 1972, 1983. *August Vollmer: Pioneer in police professionalism.* Vols. 1, 2. Berkeley: University of California, The Bancroft Library.

Rose, Nikolas. 1992. Engineering the human soul: Analyzing psychological expertise. *Science in Context* 5 (2): 351–69.

Ross, Helen, and Marth Teghtsoonian. 2009. The curious case of Luther Trant and Weber's Law. In *Fechner Day 2009,* ed. M. A. Elliott and S. Antonijevi, 271–76. Proceedings of the 25th Annual Meeting of the International Society for Psychophysics, Galway, Ireland.

Ruckmick, Christian. 1938. The truth about the lie detector. *Journal of Applied Psychology* 22:50–58.

Saferson, Richard. 2007. *Criminalistics.* Upper Saddle River: Prentice-Hall.

Schacter, Daniel. 1999. The seven sins of memory: Insights from psychology and cognitive neuroscience. *American Psychologist* 54 (3): 182–203.

Schiebinger, Londa. 1993. *Nature's body: Gender in the making of modern science.* Boston: Beacon Press.

Schleifer, Ronald. 2009. *Intangible materialism: The body, scientific knowledge, and the power of language.* Minneapolis: University of Minnesota Press.

Schmidgen, Henning. 2005. Physics, ballistics, and psychology: A history of the chronoscope in/as context, 1845–1890. *History of Psychology* 8 (1): 46–78.

Seabury, David. 1924. *Unmasking our minds.* New York: Blue Ribbon Books.

Seed, David. 1999. *American science fiction and the Cold War: Literature and film.* Edinburgh: Edinburgh University Press.

Seltzer, Mark. 1992. *Bodies and Machines.* New York: Routledge.

Sengoopta, Chandak. 2003. *Imprint of the Raj: How fingerprinting was born in colonial India.* New York: Macmillan.

Sententia, Wrye. 2001. Brain fingerprinting: Databodies to databrains. *Journal of Cognitive Liberties* 2 (3): 31–46.

Shapin, Steven, and Simon Schaffer. 1985. *Leviathan and the air-pump: Hobbes, Boyle, and the experimental life.* Temple: Princeton University Press.

Sheridan, Susan. 2002. Words and things: Some feminist debates on culture and materialism. *Australian Feminist Studies* 17 (37): 23–30.

Shouse, Eric. 2005. Feeling, emotion, affect. *M/C Journal* 8 (6). Available at: http://journal.media-culture.org.au/0512/03-shouse.php.

Shulman, Polly. 2002 Liar liar pants on fire. *Popsci.* Posted July 24. http://www.popsci.com/scitech/article/2002-07/liar-liar-pants-fire.

Shusterman, Ronald. 1998. Ravens and writing-desks: Sokol and the two cultures. *Philosophy and Literature* 22 (1): 119–35.

Siegel, Mark. 1988. *Hugo Gernsback: Father of modern science fiction.* San Bernadino: Borgo Press.

Sinclair, Upton. 1930. *Mental radio.* 2nd ed. New York: Albert and Charles Boni.

Smith, Crosbie. 1998. *The science of energy: A cultural history of energy physics in Victorian Britain.* Chicago: University of Chicago Press.

Smith, Erin. 2000. *Hard-boiled: Working-class readers and pulp magazines*. Philadelphia: Temple University Press.

Smith, Robert, and Brian Wynne. 1989. *Expert evidence: Interpreting science and the law*. London: Routledge.

Smith, Roger. 1992. *Inhibition: History and meaning in the sciences of mind and brain*. Berkeley: University of California Press.

Society for Neuroscience. 2002. Understanding human behavior: Deception. *Brain Waves*. http://apu.sfn.org/content/Publications/BrainWaves/PastIssues/2002spring/. Accessed on January 27, 2007.

Solovitch, Sara. 2004. Mind reader. *Legal Affairs: The Magazine at the Intersection of Law and Life*. http://www.legalaffairs.org/issues/November-December-2004/story_solovitch_novdec04.html.

Sommers, Asher van A. 1930. Science, the police—and the criminal. *Scientific Detective Monthly*, April.

Spence, Sean, Tom Farrow, Amy Herford, Lain Wilkinson, Ying Zheng, and Peter Woodruff. 2001. Behavioral and functional anatomical correlates of deception in humans. *Neuroreport* 12 (13): 157–64.

Spence, Sean, Mike Hunter, Tom Farrow, Russell Green, David Leung, Catherine Hughes, and Venkatasubramanian Ganesan. 2004. A cognitive neurobiological account of deception: Evidence from functional neuroimaging. *Philosophical Transactions of the Royal Society London Biological Sciences* 359 (1451): 1755–62.

Spun, Brandon. 2002. IP: Medical detection of false witness. *Insight on the News*. http://www.insightmag.com/news/2002/02/04/National/Medical.Detection.Of.False.Witness-163049.shtml. Accessed January 27, 2007.

Squier, Susan, ed. 2003. *Communities of the air: Radio century, radio culture*. Durham: Duke University Press.

Squier, Susan. 2004. *Liminal lives: Imagining the human at the frontiers of biomedicine*. Durham: Duke University Press.

Stam, Henderikus. 1998. The body's psychology and psychology's body: Disciplinary and extra-disciplinary examinations. In *The body and psychology*, ed. Henderikus Stam. London: Sage Publications.

Star, Susan Leigh. 1989. *Regions of the mind: Brain research and the quest for scientific certainty*. Stanford: Stanford University Press.

Stearns, Peter N. 1994. *American cool: Constructing a twentieth-century emotional style*. New York: New York University Press.

Stearns, Peter N., and Jan Lewis, eds. 1998. *An emotional history of the U.S.* New York: New York University Press.

Steiner, George. 1975. *After Babel: Aspects of language and translation*. Oxford: Oxford University Press.

Stenbrunner, Chris, and Otto Penzler. 1984. *Encyclopedia of mystery and detection*. New York: Harcourt.

Stevenson, Robert Louis. 1906. The body snatchers. In *The works of Robert Louis Stevenson*, edited by Charles Curtis Bigelow and Temple Scott. New York: Davos Press.

Streitfeld, David, and Charles Piller. 2002. Big Brother finds ally in once-wary high tech. *Los Angeles Times*. http://www.chicagotribune.com/technology/local/profiles/chi-020119techterror,1,7845118.story.

Stucchi, Natale. 1996. Seeing and thinking: Vittorio Benussi and the Graz School. *Axiomathes* 1–2:137–72.

Thiher, Allen. 2005. *Fiction refracts science: Modernist writers from Proust to Borges.* Columbia: University of Missouri Press.

Thomas, Ronald. 1999. *Detective fiction and the rise of forensic science.* Cambridge: Cambridge University Press.

Thompson, Clive. 2001. The lie detector that scans your brain. *New York Times Magazine,* December 9.

Tran, Trinh. 2002. Truth and deception—the brain tells all. *Penn Current.* http://www.upenn.edu/pennnews/current/2002/041102/research.html.

Trovillo, Paul V. 1939. A history of lie detection (part 1). *Journal of Criminal Law and Criminology* 29 (6): 848–81.

Trovillo, Paul V. 1939. A history of lie detection (part 2). *Journal of Criminal Law and criminology* 30 (1): 104–19.

Unger, Debi, and Irwin Unger. 2005. *The Guggenheims.* New York: HarperCollins.

Vandenbosch, Katrien, Bruno Verschuere, Geert Crombez, and Armand De Clercq. 2008. The validity of finger pulse line length for the detection of concealed information. *International Journal of Psychophysiology* 71 (2): 118–23.

Vedantam, Shankar. 2001. The polygraph test meets its match: Researchers find brain scans can be powerful tool in detecting lies. *Washington Post,* A-2.

Vincent, Harl. 1929. Barton's island. *Amazing Stories.* August.

Vollmer, August. 1937. *Crime, crooks, and cops.* New York: Funk and Wagnalls.

Waldby, Catherine. 2000. *The Visible Human Project: Informatic bodies and posthuman medicine.* London: Routledge.

Walker, Samuel. 1977. *A critical history of police reform: The emergence of professionalism.* Lexington: D. C. Heath.

Walker, Samuel. 1998. *Popular justice: A history of American criminal justice.* New York: Oxford University Press.

Warcollier, René. 1963. *Mind to mind.* Trans. J. Gridley, E. P. Matthews, and Herma Briffault. Ed. E. Schwartz. New York: Collier. Original edition, 1948.

Ward, Steven. 2002. *Modernizing the mind: Psychological knowledge and the remaking of society.* Westport: Praeger.

Watson, James, and Francis Crick. 1953. Molecular structure of nucleic acids—a structure for deoxyribose nucleic acid. *Nature* 171 (4356): 737–38.

Wegenstein, Bernadette. 2002. Getting under the skin, or, how faces have become obsolete. *Configurations* 10 (2): 221–59.

Weinbaum, Stanley. 1974. *A Martian odyssey and other science fiction tales: The collected short stories of Stanley G. Weinbaum.* Westport: Hyperion Press.

Wertheimer, Max, and Julius Klein. 1904. Psychologische Tatbestandsdiagnostik. Ideen zu psychologish-experimentellen Methoden zum Zweck der Festellung der Anteilnahme eines Menschen an einem Tatbestande. *Archiv fur Kriminalanthropologie und Kriminalistik* 15:72–113.

Wertheimer, Michael, D. Brett King, Mark A. Peckler, Scott Raney, and Roddy W. Schaef. 1992. Carl Jung and Max Wertheimer on a priority issue. *Journal of the History of the Behavioral Sciences* 28 (1): 45–56.

Wigmore, John H. 1909. Professor Münsterberg and the psychology of testimony: Being a report of the case of Cokestone v. Münsterberg. *Illinois Law Review* 3:399.

Wilcox, Daniel, and Daniel E. Sosnowski. 2005. Polygraph examination of British sexual offenders: A pilot study on sexual history disclosure testing. *Journal of Sexual Aggression* 11 (1): 3–25.

Wilson, Elizabeth. 1998. *Neural geographies: Feminism and the microstructure of cognition.* New York: Routledge.

Wilson, Harold. 1970. McClure's Magazine *and the muckrakers.* Princeton: Princeton University Press.

Wired Science. 2007. Advertisement. *Real Simple Magazine.* October.

Witchalls, Clint. 2004. Murder in mind. *Guardian.* March. http://www.guardian.co.uk/crime/article/0,2763,1176912,00.html.

Wolfe, Amy, David Bjornstad, Milton Russell, and Nichole Kerchner. 2002. A framework for analyzing dialogues over the acceptability of controversial technologies. *Science, Technology, and Human Values* 27 (1): 134–59.

Wolpe, Paul, Kenneth Foster, and Daniel Langleben. 2005. Emerging neurotechnologies for lie-detection: Promises and perils. *American Journal of Bioethics* 5 (2): 39–49.

Wolpe, Paul Root, and Daniel Langleben. 2008. Breakthrough ideas for 2008: Lies, damn lies, and lie detectors. *Harvard Business Review* 86 (2): 17–45, 25.

Would you dare take these tests? 1938. *Look Magazine,* December 6.

Wundt, Wilhelm. 1880. *Grundzfige der physiologischen Psychologie.* 2nd ed. Leipzig: Engelmann.

Yerkes, Robert M. 1923. Psychological work of the National Research Council. *Annals of the American Academy of Political and Social Science* 110:172–78.

Yerkes, Robert M., and C. S. Berry. 1909. The association reaction method of mental diagnosis. *American Journal of Psychology* 20:22–27.

# Index

191

Edwards Brothers Inc.
Ann Arbor MI. USA
March 23, 2011